Application
of
Machine Learning
for
Anaerobic Digestion 1st Ed

Application of Machine Learning for Anaerobic Digestion

초판인쇄 2025년 9월 25일 **지은이** Choi Sujin Jannat Md Abu Hanifa Hwang Seokhwan **펴낸이** 김영태 **펴낸곳** 도서출판 한비CO **출판등록** 2006년 1월 4일 제 25100-2006-1호 **주소** 700-442 대구시 중구 남산2동 938-8번지 미래빌딩 3층 301호 **전화** 053)252-0155 **팩스** 053)252-0156 **홈페이지** http://hanbimh.co.kr **이메일** kskhb9933@hanmail.net

ISBN 9791164871711
값 100,000원
*잘못된 책은 교환해 드립니다.
*저자와의 협의로 인지는 생략합니다.

Application
of
Machine Learning
for
Anaerobic Digestion

Choi Sujin Jannat Md Abu Hanifa Hwang Seokhwan

Application of Machine Learning for Anaerobic Digestion 1st Ed.

TABLE OF CONTENTS

PREFACE -- ii

TABLE OF CONTENTS ---------------------------------- iii

Chapter 1. Anaerobic Digestion ---------------------- 1

    1.1 Introduction ------------------------------------ 1

    1.2 Biochemistry and Microbiology -------------- 3

        1.2.1 Hydrolysis ----------------------------------- 3

        1.2.2 Acidogenesis ------------------------------ 4

        1.2.3 Acetogenesis ------------------------------ 4

        1.2.4 Methanogenesis -------------------------- 5

    1.3 Process Control of Anaerobic Digestion ---- 6

Chapter 2. Statistical Methods in Anaerobic Digestion
-- **11**

    2.1 Multiple regression ------------------------ 11

        2.1.1 Fundamentals of multiple regression -- 12

        2.1.2 Response surface methodology -------- 15

            2.1.2.1 Design of experiments in AD ----- 16

            2.1.2.2 Central composite design --------- 17

2.1.2.3 Full factorial and fractional factorial designs -- 22
2.1.3 Application of RSM in AD processes -- 26
2.2 Multivariate analysis techniques ------------ 29
2.2.1 Overview of multivariate methods in AD -- 29
2.2.2 Principle component analysis (PCA) --- 32
    2.2.2.1 Mathematical foundation ---------- 33
    2.2.2.2 Application of PCA in AD --------- 34
    2.2.2.3 Interpretation of PCA results ----- 35
    2.2.2.4 Interpretation of key parameters - 36
2.2.3 Redundancy analysis -------------------- 39
    2.2.3.1 Linking environmental parameters to microbial communities ----------------------- 41
2.2.4 Principle Coordinate analysis ----------- 48
    2.2.4.1 Applications in sample comparison and clustering ---------------------------------- 51
2.2.5 Non-metric multidimensional scaling -- 58
    2.2.5.1 Concept and application in microbial community analysis ---------------------------- 61

2.2.6 Canonical Correspondence Analysis ---- 68
    2.2.6.1 Correlating microbial communities and process performance --------------------- 71

Chapter 3. Kinetics of Microbial Growth and Substrate Utilization -- 78

3.1 General concepts of biokinetics in AD ---- 78
    3.1.1 Definition and importance -------------- 82
    3.1.2 Microbial growth curve ----------------- 83

3.2 Importance of Biokinetics in AD Process Contro --- 89

3.3 Fundamental Kinetic Models in AD -------- 96
    3.3.1 Monod Model ------------------------- 97
    3.3.2 Haldane Model ------------------------ 99
    3.3.3 Contois Model --------------------------100
    3.3.4 Lotka-Volterra Model ------------------102
        3.3.4.1 Case studies: Interaction modeling --108

3.4 Fundamental biokinetic equations --------- 111
    3.4.1 Mass balance equation development - 113
    3.4.2 Analytical solutions -------------------- 115

    3.4.2.1 Solution for batch reactors ------ 115

    3.4.2.2 Solution for CSTRs --------------- 117

  3.4.3 Numerical solutions -------------------- 119

    3.4.3.1 4th order Runge-Kutta Method -- 119

    3.4.3.2 Solution for batch reactors ------ 121

    3.4.3.3 Solution for CSTRs --------------- 123

3.5 Kinetic parameter estimation -------------- 125

3.6 Applications of biokinetic modeling in performance prediction ------------------------ 128

Chapter 4. Artificial Intelligence (AI)-based methods in Anaerobic Digestion --------------------------------- 131

  4.1 General Introduction to AI ----------------- 131

    4.1.1 Basics of AI ----------------------------- 134

    4.1.2 AI Approach in Anaerobic Digestion - 136

    4.1.3 Common AI Tasks in Anaerobic Digestion systems --------------------------------------- 138

      4.1.3.1 Prediction ------------------------- 140

      4.1.3.2 Classification---------------------- 141

      4.1.3.3 Clustering --------------------------141

4.1.3.4 Anomaly Detection --------------- 142
4.2 Description of Model Architectures ------- 144
    4.2.1 Machine Learning Algorithms --------- 145
        4.2.1.1 Linear Regression ---------------- 147
        4.2.1.2 Support Vector Machine (SVM) - 151
        4.2.1.3 Naïe Bayes ----------------------- 156
        4.2.1.4 Decision Trees ------------------- 161
        4.2.1.5 Ensemble Learning Models ------ 166
    4.2.2 Neural Network Architectures --------- 174
        4.2.2.1 Multi-layer Perceptrons (MLP) --- 175
        4.2.2.2 Convolutional Neural Networks (CNN)
--182
        4.2.2.3 Recurrent Neural Networks (RNN) - 194
        4.2.2.4 Transformer ---------------------- 206
4.3 Training and Validation of Models ------- 219
    4.3.1 Data Splitting -------------------------- 222
    4.3.2 Learning Objectives -------------------- 223
    4.3.3 Model Optimization ------------------- 225
    4.3.4 Validation and Evaluation of Models - 229

Chapter 5. Application of AI in Anaerobic Digestion research 234

- 5.1 Physico-Chemical Data 234
  - 5.1.1 Reactor Configuration 235
  - 5.1.2 Substrate Characteristics 236
  - 5.1.3 Effluent Characteristics 238
  - 5.1.4 Gas Phase Data 239
  - 5.1.5 Applications in AI Modeling 241
- 5.2 Qualitative and Quantitative Microbial Data - 243
  - 5.2.1 Qualitative Microbial Data 244
    - 5.2.1.1 Community Structure 244
    - 5.2.1.2 Diversity Indices 245
    - 5.2.1.3 Applications in AI Modeling 246
  - 5.2.2 Quantitative Microbial Data 248
    - 5.2.2.1 Microbial Quantification using qPCR - 249
    - 5.2.2.2 Applications in AI Modeling 249
  - 5.2.3 Integration of Qualitative and Quantitative Data - 250
- 5.3 Image Data 252
  - 5.3.1 Spectroscopy Image 254
    - 5.3.1.1 Application in Influent Analysis 254

    5.3.1.2 Application in Effluent Analysis - 254

    5.3.1.3 Applications in AI Modeling ----- 255

  5.3.2 Microscopy Image --------------------- 257

    5.3.2.1 Biofilm and Granule Analysis ---- 258

    5.3.2.2 Microbial Community Imaging -- 258

    5.3.2.3 Applications in AI Modeling ----- 259

5.4 Time-series Data --------------------------- 261

  5.4.1 Characteristics of Time-series Data --- 262

  5.4.2 Preprocessing Requirements for AI Applications ------------------------------------- 263

  5.4.3 Applications in AI Modeling ---------- 264

REFERENCES ------------------------------------ 266

Chapter 1. Anaerobic Digestion

1.1 Introduction

Anaerobic digestion (AD) is a biological process in which microorganisms break down biodegradable organic matter in the absence of oxygen, leading to the production of biogas—a mixture primarily composed of methane (CH_4) and carbon dioxide (CO_2). This process plays a critical role in sustainable waste management and renewable energy production, offering an efficient solution for the treatment of wastewater sludge, agricultural residues, and organic solid waste.

AD iswidely utilized in municipal wastewater treatment plants (WWTPs), industrial effluent treatment, and agricultural biogas systems due to its ability to reduce waste volume, mitigate greenhouse gas emissions, and generate renewable energy. The process aligns with circular economy principles, promoting resource recovery through energy generation and nutrient recycling in the form of digestate, which can be used as a soil amendment.

Historically, AD has been employed for over a century, but recent advances in microbiological understanding, process control, and reactor engineering have significantly enhanced its efficiency and applicability. Current research efforts focus on improving methane yield, optimizing microbial consortia, integrating AD with other technologies (e.g., membrane filtration, thermal hydrolysis, and co-digestion strategies), and mitigating inhibitory factors to ensure stable operation.

1.2 Biochemistry and Microbiology

AD is a cascade of interdependent biochemical reactions mediated by syntrophic bacteria and methanogenic archaea (Fig. 1). The four-stage process—hydrolysis, acidogenesis, acetogenesis, and methanogenesis—requires precise thermodynamic and kinetic coordination.

1.2.1. Hydrolysis

Complex organic polymers (cellulose, proteins, lipids) are hydrolyzed into monomers (glucose, amino acids, fatty acids) via extracellular enzymes (e.g., cellulases, proteases):

$$\text{Cellulose } (C_6H_{10}O_5)_n + nH_2O \xrightarrow{\text{Cellulase}} nC_6H_{12}O_6 \quad (\Delta G° = -4\,\text{kJ/mol})$$

Hydrolytic bacteria (e.g., *Clostridium*, *Bacteroides*) dominate this stage, but kinetics are often limited by lignin-rich substrates (Tsapekos et al., 2017). Pretreatment strategies (thermal, enzymatic) enhance hydrolysis rates by disrupting recalcitrant structures (Zheng et al., 2014).

1.2.2 Acidogenesis

Monosaccharides undergo fermentation to volatile fatty acids (VFAs), alcohols, H_2, and CO_2:

$$C_6H_{12}O_6 \rightarrow 2CH_3CH_2OH + 2CO_2 \quad (\Delta G^\circ = -235\,kJ/mol)$$

$$C_6H_{12}O_6 + 2H_2O \rightarrow 2CH_3COOH + 2CO_2 + 4H_2 \quad (\Delta G^\circ = -206\,kJ/mol)$$

Dominant genera (*Enterobacter*, *Lactobacillus*) thrive at pH 5.0–6.0, but excessive VFA accumulation (<4 g/L) inhibits methanogens (Chen et al., 2008).

1.2.3 Acetogenesis

Long-chain VFAs (e.g., propionate, butyrate) are oxidized to acetate, H_2, and CO_2 by obligate hydrogen-producing acetogens (OHPA):

$$CH_3CH_2COOH + 2H_2O \rightarrow CH_3COOH + CO_2 + 3H_2 \quad (\Delta G^\circ = +76\,kJ/mol)$$

This endergonic reaction is feasible only when H_2 partial pressure (<10^{-4} atm) is maintained via hydrogenotrophic methanogens (Stams & Plugge, 2009). *Syntrophobacter* and *Syntrophomonas* are keystone species in this syntrophy.

1.2.4 Methanogenesis

Methanogenic archaea utilize three pathways:

a) Acetoclastic (dominant in mesophilic systems):

$$CH_3COOH \rightarrow CH_4 + CO_2 \quad (\Delta G° = -31\,kJ/mol)$$

b) Hydrogenotrophic (critical under thermophilic/high-NH$_3$ conditions):

$$CO_2 + 4H_2 \rightarrow CH_4 + 2H_2O \quad (\Delta G° = -131\,kJ/mol)$$

c) Methylotrophic (rare, using methanol/methylamines).

Methanosarcina spp. exhibit metabolic flexibility, while *Methanobacterium* are obligate hydrogenotrophs (Angelidaki et al., 2011). Ammonia (NH$_3$ > 200 mg/L) disrupts ion gradients in methanogens, while sulfide

(H_2S > 100 mg/L) inhibits coenzyme M reductase (Fotidis et al., 2014).

Metagenomic studies reveal *Methanoculleus* dominance under high-ammonia stress, while bioaugmentation with *Methanothrix* enhances acetate utilization (Treu et al., 2016).

1.3 Process Control of Anaerobic Digestion

To ensure the stable operation of anaerobic digestion systems, various operational parameters and control strategies must be carefully monitored and optimized.

1) pH and Alkalinity

The optimal pH range for AD typically falls between 6.8 and 7.5, with methanogens being highly sensitive to acidic conditions. Alkalinity serves as a buffer against acidification, preventing process failure due to excessive VFA accumulation.

2) Temperature Control

AD systems operate under three primary temperature regimes:
- Psychrophilic (10–20°C): Low-energy consumption but slow kinetics.
- Mesophilic (35–40°C): Most common due to process stability and energy efficiency.
- Thermophilic (50–60°C): Higher reaction rates and pathogen removal but greater sensitivity to operational fluctuations.

3) Organic Loading Rate (OLR)

OLR, expressed as kg $COD/m^3 \cdot d$, represents the amount of organic material introduced per unit reactor volume per day. Excessive loading can lead to acid accumulation and system failure, while insufficient loading results in underutilization of reactor capacity.

4) Hydraulic Retention Time (HRT) and Solids Retention Time (SRT)

HRT and SRT dictate the residence time of substrates and microbial biomass within the reactor. Long SRTs favor methanogenic growth, while short HRTs can lead to washout of slow-growing microorganisms.

5) Inhibitory Compounds

AD performance can be hindered by various inhibitory substances, including ammonia (NH_4^+/NH_3), sulfides (H_2S), heavy metals, and toxic organic compounds. Ammonia inhibition, for instance, is a major challenge in the digestion of protein-rich substrates:

$$NH_4^+ + OH^- \rightleftharpoons NH_3 + H_2O$$

Control strategies, such as feedstock dilution, co-digestion, and pH adjustment, are employed to mitigate inhibitory effects and maintain system stability.

6) Reactor Configurations

Several reactor types are utilized to optimize AD performance, including:
- Completely Stirred Tank Reactor (CSTR): Widely used in WWTPs for sludge digestion.
- Up flow Anaerobic Sludge Blanket (UASB): High-rate reactor with excellent biomass retention.
- Anaerobic Membrane Bioreactor (AnMBR): Integrates membrane filtration to enhance effluent quality and biomass retention.

7) Process Monitoring and Automation

Advanced monitoring tools, such as online sensors for biogas composition, volatile fatty acids, and real-time microbial activity tracking, enable improved process control. Machine learning and artificial intelligence (AI)-based predictive models are increasingly being integrated into AD operations to enhance efficiency and resilience.

8) Conclusion

Anaerobic digestion is a vital technology for sustainable waste management and bioenergy production. Understanding its biochemical pathways, microbial ecology, and process control mechanisms is essential for optimizing performance and ensuring long-term operational stability. Continued research and technological advancements will further enhance its efficiency, scalability, and integration into circular bio economy frameworks.

Chapter 2. Statistical Methods in Anaerobic Digestion

2.1 Multiple regression

Anaerobic digestion (AD) is a highly complex biochemical process involving the coordinated activity of diverse microbial consortia under tightly regulated environmental and operational conditions (Demirel et al., 2008). Due to the multiplicity of factors influencing system behavior and performance, the application of statistical modeling becomes indispensable. Among these, multiple regression analysis stands out as a fundamental approachto understand the relationship between a response variable and multiple explanatory variables. In recent years, it has emerged as a valuable tool in optimizing and predicting the performance of AD processes, aiding in the rational design and operation ofdigesters.

Multiple regression extends simple linear regression by incorporating two or more independent variables. In the context of AD, this might include temperature, pH, organic loading rate (OLR), hydraulic retention time (HRT), substrate composition, and microbial parameters, all influencing outputs such as methane production, volatile solids reduction, and effluent quality (Mata-Alvarez et al., 2011). By establishing a statistical model that relates these inputs to the output, researchers and practitioners can better predict outcomes and explore the effects of parameter interactions (Xu et al., 2018).

2.1.1 Fundamentals of multiple regression

Multiple regression is a statistical technique that models the linear relationship between a dependent (response) variable and multiple independent (predictor) variables (Montgomery et al., 2021). The general form of the multiple linear regression model is:

$$Y = \beta_0 + \beta_1 X_1 + \beta_2 X_2 + \cdots + \beta_n X_n + \varepsilon \quad \ldots (2.1)$$

Where,

Y is the dependent variable (e.g., methane yield),

$X_1, X_2, ..., X_n$ are the independent variables (e.g., temperature, pH, HRT),

β_0 is the intercept,

$\beta_1, \beta_2, ..., \beta_n$ are the regression coefficients,

ε is the error term representing the deviation of the observed values from the predicted values.

For multiple regression to be valid, several assumptions must be met (Montgomery et al., 2021; Neter et al., 1996).

- Linearity: The relationship between the independent variables and the dependent variable is linear.
- Independence: Observations must be independent of each other.
- Homoscedasticity: Constant variance of the errors across all levels of the independent variables.
- Normality: The residuals (errors) should be normally distributed.
- No multicollinearity: The independent variables should not be too highly correlated with one another.

The performance of a multiple regression model is typically assessed using (Draper, 1998):

- R-squared (R^2): Proportion of variance in the dependent variable explained by the independent variables.
- Adjusted R^2: Adjusted for the number of predictors in the model, useful for comparing models.
- F-statistic: Tests the overall significance of the model.
- p-values: Indicates the statistical significance of each predictor.
- Residual plots: Help assess whether assumptions have been met.

In AD systems, multiple regression is used to predict methane yield, assess the effect of varying OLR and HRT, quantify inhibition effects (e.g., ammonia), and model microbial responses to operational changes (Xu et al., 2018). The models are built using experimental or operational data and are validated using part of the dataset or through external validation.

2.1.2 Response surface methodology (RSM)

Response Surface Methodology (RSM) is an advanced form of regression modeling that incorporates both linear and quadratic terms, enabling the modeling of curvature and interactions among variables. It is particularly useful in optimization studies where the goal is to find the best combination of variables that maximize or minimize a response (Myers et al., 2016). The general second-order RSM model is:

$$Y = \beta_0 + \Sigma\beta_i X_i + \Sigma\beta_{ii} X_i^2 + \Sigma\beta_{ij} X_i X_j + \varepsilon \ldots\ldots\ldots\ldots\ldots\ldots\ldots\ldots\ldots(2.2)$$

Where β_i represents linear effects, β_{ii} represents quadratic effects, and β_{ij} represents interaction effects. RSM provides detailed insights into the interactions between variables and the nonlinear nature of AD systems (Jannat et al., 2021; Yasin et al., 2013). It allows the development of predictive models and response surfaces that can be visualized through contour plots and 3D surface plots.

2.1.2.1 Design of experiments in AD

Design of Experiments (DoE) is a structured approach for planning experiments to ensure that data obtained can be analyzed to yield valid and objective conclusions. In the context of AD, DoE is used to efficiently study the effects of multiple parameters and their interactions on responses like biogas production and process stability (El-Mashad & Zhang, 2010). Typical steps include:
- Selection of variables: Based on prior knowledge or preliminary studies.
- Defining response variables: Common responses include methane yield, COD removal, or pH stability.
- Choosing an experimental design: Such as factorial, fractional factorial, or central composite designs.

- Execution of experiments: Systematic variation of inputs.
- Statistical analysis: Fitting regression models, analyzing variance (ANOVA).

The strength of DoE in AD research lies in its efficiency—obtaining maximum information with minimum experiments. It helps in identifying significant factors, quantifying their effects, and guiding further experimentation.

2.1.2.2 Central composite design

Central Composite Design (CCD) is a widely employed experimental design under the framework of RSM. It is particularly suitable for developing second-order polynomial models that can represent both the linear and quadratic effects of independent variables, as well astheir interactions (Montgomery et al., 2021; Myers et al., 2016). CCD is highly efficient for optimization in systems like anaerobic digestion,

where the behavior is governed by multiple interdependent factors. Structure of CCD typically consists of three sets of experimental points:
- Factorial points: Representing a full or fractional factorial design, which covers all combinations of high and low levels of factors (coded as +1 and -1).
- Axial (star) points: These are added to estimate curvature. Each factor is varied beyond its high and low levels to capture quadratic behavior. The axial points are placed at a distance α from the center point.
- Center points: Replicated runs at the center of the design space, which help assess experimental error and detectnon-linearity.

The second-order model fitted using CCD is expressed as:

$$Y = \beta_0 + \Sigma\beta_i X_i + \Sigma\beta_{ii} X_i^2 + \Sigma\beta_{ij} X_i X_j + \varepsilon \quad \ldots\ldots\ldots\ldots\ldots\ldots\ldots(2.3)$$

Here, β_0 is the intercept, β_i are the linear coefficients, β_{ii} are the quadratic coefficients, and β_{ij} represent the interaction effects between the variables X_i and X_j.

Advantages of CCD in AD Research:
- Efficiency: CCD requires fewer experimental runs than full three-level factorial designs, making it cost-effective and time-saving.
- Modeling flexibility: Capable of modeling complex, nonlinear interactions between multiple process parameters.
- Error estimation: The use of center points allows for estimation of pure error, aiding in model validation.
- Scalability: CCD is adaptable to any number of process variables, which is especially beneficial in multifactorial AD processes.

In AD research, CCD has been applied to optimize parameters such as:

- Substrate concentration and composition: Determining the optimal feed mix for co-digestion.
- pH and temperature: Key factors influencing microbial activity.
- Hydraulic Retention Time (HRT) and Organic Loading Rate (OLR): Influencing reactor stability and performance.
- Inhibitor concentrations: Identifying thresholds for substances like ammonia or sulfides.

After conducting experiments, regression analysis is used to fit the second-order model. The model is evaluated using ANOVA to determine the significance of each term. Non-significant terms may be removed to simplify the model. Response surface and contour plots are generated to visualize interactions. Model Validation is performed by: Comparing predicted responses with actual outcomes, using diagnostic plots (residual vs. fitted, normal probability plots), and performing additional experiments near the predicted optimum to verify model accuracy. Axial points may extend beyond feasible operational ranges. Assumes quadratic behavior, not suitable if the true system behavior is more complex. Requires proper coding of variables and transformation of responses (ifneeded) for linearity.

CCD provides a systematic, statistically robust, and efficient approach to experiment planning and optimization in anaerobic digestion research. It has proven instrumental in enhancing understanding of process interactions, identifying optimal operating conditions, and reducing the cost and time of experimentation (El-Mashad & Zhang, 2010). As AD systems become more sophisticated with the integration of multi-substrate feedstock's and variable operational conditions, the utility of CCD will only grow. Its ability to model complex relationships and predict system behavior makes it a critical tool for researchers and practitioners striving for higher efficiency and sustainability in anaerobic processes (Li et al., 2011).

2.1.2.3 Full factorial and fractional factorial designs

Factorial designs, both full and fractional, are foundational elements in experimental design, particularly suited for systems such as anaerobic digestion where multiple input parameters may interact in complex ways. These designs are used to systematically investigate the effects of two or more factors by varying them simultaneously rather than one at a time, thus providing insights into both main effects and interaction effects (Myers et al., 2016).

A full factorial design examines all possible combinations of the levels of factors. For example, a 2^k factorial design involves k factors, each at two levels (usually coded as -1 and +1), and requires 2^k experimental runs. In the case of three factors (e.g., temperature, pH, and OLR), a full factorial design would require $2^3 = 8$ runs. Advantages: Captures all interactions between factors, allows full estimation of the model including higher-order terms, ideal for small numbers of factors (2-5) due to feasibility. Limitations: becomes exponentially resource-intensive as the number of factors increases, might be impractical for large-scale or resource-constrained experiments. Full factorial designs are commonly used in AD research when the number of experimental variables is small. For instance, a study examining the influence of C/N ratio, inoculum-to-substrate ratio (ISR), and temperature on methane yield might use a 2^3 full factorial design to identify significant variables and their interactions. The resultant data can then be analyzed using ANOVA to determine the statistical significance of effects.

To reduce the number of experimental runs required by full factorial designs, fractional factorial designs (FFDs) are used. These designs include only a subset (fraction) of the full factorial runs and are particularly beneficial when many factors are being studied, but only a few are expected to be significant. A fractional factorial design is denoted as $2^{(k-p)}$, where: k = number of factors, p = fraction of the full factorial design being used. For example, a $2^{(5-2)}$ fractional factorial design involves 5 factors but only 8 runs instead of 32 in a full design. Key Concepts: Resolution: Indicates the degree of confounding between main effects and interactions. A resolution III design may confound main effects with two-factor interactions, whereas a resolution V design keeps main effects and two-factor interactions unconfounded. Aliasing: In fractional designs, some effects are indistinguishable (aliased) from others due to reduced run numbers. Understanding alias structures is critical. Advantages: Reduces number of runs while still estimating main effects. Ideal for screening studies to identify key factors. Limitations: Potential for confounding and aliasing effects. Less suitable for detailed optimization unless followed by higher-resolution design. Application in AD: In AD systems, fractional factorial designs are widely used in the preliminary stages of experimentation. They are effective in screening studies to identify critical parameters from a larger pool. For example, in the optimization of co-digestion mixtures, a researcher may investigate up to seven variables (e.g., carbon source, nitrogen content, mixing ratio, temperature, pH, particle size, inoculum type) using a $2^{(7-3)}$ design with only 16 runs, drastically saving time and resources.

Data from factorial and fractional factorial designs are analyzed using ANOVA to determine significant factors (Montgomery et al., 2021). Main effects plots, interaction plots, and Pareto charts are used to visually represent and interpret the results. These tools help prioritize which variables and interactions should be included in further modeling efforts. Both full factorial and fractional factorial designs serve as vital tools in anaerobic digestion research. While full factorial designs offer comprehensive insights, their practicality diminishes with an increasing number of factors. In contrast, fractional factorial designs provide a practical means of screening and understanding complex systems efficiently. The strategic use of these designs enhances the scientific rigor, cost-effectiveness, and interpretability of AD process optimization efforts. They often serve asthe preliminary step before advancing to more intricate designs like CCD or Box-Behnken designs in the framework of RSM (Myers et al., 2016).

2.1.3 Application of RSM in AD process

The practical implementation of Response Surface Methodology (RSM) in anaerobic digestion (AD) processes has gained widespread attention due to its ability to optimize complex and nonlinear biological systems. Anaerobic digestion is influenced by a multitude of physicochemical and operational parameters—such as substrate composition, inoculum characteristics, temperature, pH, hydraulic retention time (HRT), organic loading rate (OLR), carbon-to-nitrogen (C/N) ratio, and mixing intensity—which interact in nonlinear ways (Li et al., 2011). RSM offers a systematic and statistically grounded approach to model these interactions and optimize system outputs, primarily biogas or methane yield.

In RSM, after selecting the appropriate experimental design (e.g., central composite design or Box-Behnken design), experiments are conducted, and a second-order polynomial model is fitted to the response data (Ferreira et al., 2007). The model includes linear terms, quadratic terms, and interaction terms that capture the effects of individual parameters and their combinations on the response variable. Once the model is established, response surfaces and contour plots are generated to visualize how the response varies with changes in the predictor variables. These plots are instrumental in identifying optimal conditions and understanding interactions between parameters (El-Mashad & Zhang, 2010; Myers et al., 2016).

2.2 Multivariate analysis techniques
2.2.1 Overview of multivariate methods in AD

Anaerobic digestion (AD) is inherently a complex and dynamic biotechnological process governed by the simultaneous interaction of numerous physical, chemical, and biological variables (Appels et al., 2008; Batstone et al., 2002). These include operational parameters such as temperature, pH, organic loading rate (OLR), hydraulic retention time (HRT), and mixing intensity, as well as system outputs like biogas composition, methane yield, volatile solids (VS) reduction, volatile fatty acids (VFAs), ammonia concentration, and microbial community structure (Li et al., 2011). Traditional univariate and bivariate statistical methods are often inadequate for capturing the full picture of how these interrelated variables co-vary, interact, and collectively influence process performance and stability. As AD technology continues to evolve toward higher efficiency and tighter process control, the need for robust, data-driven tools capable of interpreting large, multidimensional datasets has become increasingly urgent (Zamanzadeh et al., 2017).

Multivariate analysis (MVA) techniques provide the statistical foundation to tackle this challenge. MVA refers to a family of statistical methods designed to analyze datasets containing multiple variables simultaneously, revealing patterns and relationships that cannot be discerned through simple pairwise comparisons (Jolliffe et al., 2016). In anaerobic digestion research and operation, multivariate methods have found broad application in:

- Process monitoring and fault detection: Identifying abnormal operating conditions based on the combined behavior of multiple variables.
- Performance optimization: Linking input variables (feedstock properties, operational parameters) with outputs (biogas yield, effluent quality).
- Microbial ecology studies: Unraveling relationships between community structure and process performance.

- Predictive modeling: Developing data-driven models for forecasting biogas production or system failures.
- Experimental design and factor screening: Determining which variables significantly impact target responses.

The rise of high-frequency sensor networks, online analyzers, and high-throughput sequencing has vastly increased the amount and complexity of data available from AD systems (Vanwonterghem et al., 2014). Multivariate analysis techniques such as Principal Component Analysis (PCA), Partial Least Squares Regression (PLSR), Cluster Analysis, and Discriminant Analysis have therefore become indispensable tools for both researchers and practitioners (Jolliffe et al., 2016). Among these, Principal Component Analysis is particularly prominent because of its versatility in data exploration, dimensionality reduction, pattern recognition, and variable selection.

2.2.2 Principle component analysis

Principal Component Analysis (PCA) is a statistical technique used to simplify complex datasets by transforming them into a new set of uncorrelated variables called principal components. Each principal component is a linear combination of the original variables, constructed in such a way that it captures as much of the total variance in the data as possible. In practical terms, PCA reduces a high-dimensional dataset into a smaller number of dimensions while retaining the most critical information (Lever et al., 2017). This makes it possible to visualize and interpret patterns, trends, and outliers that would otherwise be obscuredin raw data tables. PCA is an unsupervised method, meaning it does not require predefined labels or classes. It is exploratory in nature and particularly useful in the early stages of AD research when the aim is to uncover hidden structures in the data (Vanwonterghem et al., 2014; Zamanzadeh et al., 2017).

2.2.2.1 Mathematical foundation

At its core, PCA involves:

I. Standardizing the data: Since variables in AD datasets often have different units (e.g., °C, mg/L, m³/d), the data are standardized (mean-centered and scaled) so that each variable contributes equally.

II. Calculating the covariance matrix: This matrix describes how pairs of variables vary together.

III. Computing eigenvalues and eigenvectors: Eigenvectors determine the directions of the new principal component axes, while eigenvalues indicate how much variance each principal component explains.

IV. Transforming the original data: The original variables are projected onto the new axes, producing principal component scores.

Typically, only the first few principal components are needed to explain most of the variance in the dataset.

2.2.2.2 Application of PCA in AD

PCA has been widely used in anaerobic digestion studies for:

- Monitoring process stability: By analyzing time-series data on pH, alkalinity, VFAs, biogas yield, and other parameters simultaneously, PCA can detect unusual operating states that deviate from the typical process behavior (Zamanzadeh et al., 2016).
- Identifying key variables: PCA helps identify which variables contribute most to system variance, thus highlighting the critical factors influencing performance (Ward et al., 2008).
- Microbial community profiling: When applied to high-dimensional sequencing data (e.g., 16S rRNA gene profiles), PCA can reveal correlations between microbial community shifts and changes in process outputs (Westerholm, 2012).
- Comparing operating regimes: PCA makes it possible to visualize and distinguish clusters

representing different operating phases, feedstock's, or treatment conditions.
- Outlier detection: Samples or operating days that deviate significantly from the main data cloud in PCA plots may indicate measurement errors, sensor drift, or abnormal process conditions.

2.2.2.3 Interpretation of PCA results

Interpreting PCA involves examining two main outputs: the scores plot and the loadings plot.
- Scores plot: Each point represents an observation (e.g., a sampling day). Clusters or trends in the scores plot reveal similarities or differences between samples.
- Loadings plot: This shows how strongly each original variable contributes to each principal component. Variables pointing in the same direction are positively correlated, while those

pointing in opposite directions are negatively correlated.

Together, these plots provide insights into the underlying structure of the dataset and help connect patterns back to physical or biological phenomena in the digester.

2.2.2.4 Interpretation of key parameters

Proper interpretation of PCA results requires careful understanding of what principal components, scores, and loadings represent in the context of anaerobic digestion.

I) Eigenvalues and Explained Variance

The eigenvalue associated with each principal component indicates how much of the total variance it accounts for. In AD studies, it is common to find that the first two or three principal components explain 70–90% of the total variance, making them sufficient for interpretation.

II) Scores

Scores are the transformed coordinates of the original data points in the new principal component space. For an AD system:
- Clustering of scores may reflect stable operation periods, similar feedstock's, or operational phases.
- Separation between clusters may indicate operational changes, feedstock shifts, or process upsets.
- Gradual drift in scores over time could suggest a gradual microbial shift or slow degradation of process stability.

III) Loadings

Loadings reveal which variables influence each principal component:
- High positive or negative loadings indicate variables that strongly impact the component.

- Variables with high loadings on the same component are correlated.
- By examining which variables dominate each component, engineers can deduce which process parameters are driving observed trends.

For example, if PCA reveals that methane yield, alkalinity, and pH have strong positive loadings on PC1, while VFAs and ammonia have strong negative loadings, this implies that PC1 represents a balance between stable methane production and acid accumulation or inhibition.

IV) Biplots

A biplot overlays scores and loadings in a single plot, providing a combined view. In AD:
- Samples located near a particular variable vector are influenced strongly by that variable.
- Clusters of samples can be interpreted in terms of which variables drive their grouping.

Multivariate analysis techniques, particularly Principal Component Analysis, have become powerful tools for understanding, controlling, and optimizing anaerobic digestion systems. By distilling complex, high-dimensional process data into a few interpretable components, PCA helps operators and researchers visualize hidden relationships, detect deviations, and link operational changes to performance trends. As AD moves towarddigitalization and real-time data analytics, PCA and related methods will continue to form the backbone of smart monitoring and advanced decision support systems (Liu et al., 2017).

2.2.3 Redundancy analysis

Redundancy Analysis (RDA) is a multivariate direct gradient analysis method that combines features of multiple regression and Principal Component Analysis (PCA) (Legendre & Legendre, 2012). It is designed to explain variation in a set of response variables (e.g., microbial community composition) by a set of explanatory variables (e.g., environmental parameters such as pH, temperature, ammonia concentration, or volatile fatty acids). While PCA is an unconstrained ordination method, finding patterns without external environmental input, RDA is a constrained ordination technique, meaning the ordination is restricted to variation that can be explained by the measured environmental parameters (Borcard et al., 2011). In AD research, RDA is particularly valuable because it enables direct quantification of how much microbial community variation is linked to changes in operational or environmental conditions. This allows researchers to disentangle the effects of feedstock changes, operational perturbations, and environmental gradients on microbial dynamics.

Conceptually, RDA is an extension of multiple linear regression into a multivariate framework. Steps:
1) Perform multiple regression of each microbial taxon abundance (response variable) on the environmental parameters (predictors).

2) Conduct PCA on the fitted values from the regression.
3) The resulting ordination axes are linear combinations of the environmental parameters that explain the greatest variance in the microbial dataset.

Mathematically:

$$Y = XB + E \quad\quad\quad\quad\quad\quad\quad\quad\quad (2.4)$$

Where, Y= Matrix of response variables (e.g., relative abundances of microbial taxa); X= Matrix of explanatory variables (e.g., pH, NH_3-N, VFA concentration); B = Regression coefficients; E = Residual matrix. The PCA is then performed on XB to produce the RDA ordination.

2.2.3.1 Linking environmental parameters to microbial communities

In the complex ecosystem of an anaerobic digester, microbial communities are inherently dynamic, responding sensitively to changes in environmental conditions (Sundberg et al., 2013; Vanwonterghem et al., 2014). The performance of AD in terms of methane yield, substrate degradation, and process stability, is tightly coupled to the structure and function of microbial consortia. Consequently, understanding how environmental parameters influence microbial dynamics is crucial for both research and operational control of AD systems. Redundancy Analysis (RDA) provides a rigorous statistical framework for exploring these relationships, enabling researchers to quantify and visualize how environmental gradients shape microbial community composition (Legendre & Legendre, 2012).

  Environmental factors such as pH, temperature, ammonium concentration, and volatile fatty acids (VFAs) are known to exert strong selective pressures on microbial populations (Rajagopal et al., 2013; Westerholm et al., 2018). For instance, methanogenic archaea exhibit narrow pH optima, typically around neutral conditions, and are sensitive to ammonia inhibition. Elevated VFAs can indicate overloading or acid accumulation, which in turn favors acidogenic bacteria while suppressing methanogens. Temperature shifts, from mesophilic to thermophilic regimes, drastically alter microbial consortia, promoting thermos-tolerant species and reducing community evenness. Each of these parameters, alone or in combination, drives shiftin microbial abundances, diversity, and functional potential. RDA allows these effects to be quantified in a multivariate context, capturing both direct and interactive influences of environmental variables on community structure (Borcard et al., 2011).

Conceptually, RDA extends multiple regression to a multivariate response framework, where each microbial taxon is modeled as a function of environmental predictors. By performing PCA on the fitted values of these regressions, RDA identifies ordination axes that are constrained to maximize the variation explained by environmental parameters. This constrained ordination is particularly advantageous in AD studies, where researchers aim not merely to describe microbial patterns, but to explicitly link them to operational and environmental conditions. The resulting RDA plots illustrate which environmental factors are most strongly associated with particular taxa, revealing ecological niches and potential functional roles (Vanwonterghem et al., 2014).

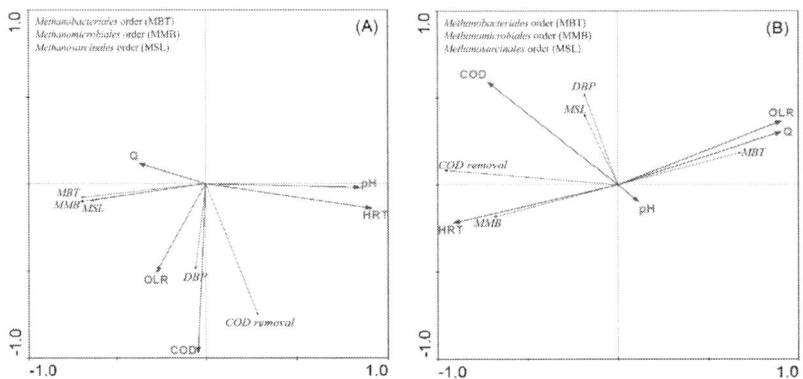

Fig 2.1 Correlation between the methanogens and environmental variables through the RDA analysis at start up stage (A) and at dynamic HRT stage (B), respectively, in farm-scale anaerobic digester (Cho et al., 2013).

A practical application of redundancy analysis (RDA) can be seen in a study investigating methanogen population behavior during the start-up of a farm-scale anaerobic digester (AD) treating swine wastewater (Cho et al., 2013). The aim was to understandhow key operational and physico-chemical parameters influenced both methanogenic community structures and digester performance. In this study, methanogenic population data (quantified via qPCR), chemical oxygen demand (COD) removal efficiency, and daily biogas production rate (DBP) were treated as dependent variables. Five operational and physico-chemical factors, hydraulic retention time (HRT), organic loading rate (OLR), influent pH, influent COD concentration, and inflow rate (Q), were selected as independent variables. RDA was performed using CANOCO 4.5 to visualize and quantify these relationships.

The RDA ordination revealed clear patterns linking environmental variables to microbial community composition and reactor performance. HRT exhibited a negative correlation with all three major methanogenic orders, suggesting that extremely long HRTs (>100 days) were suboptimal for methanogenic growth. At prolonged HRTs, dilution rates fell below the optimal range, limiting substrate availability and thus suppressing methanogen populations. Conversely, a 50-day HRT provided conditions more favorable for methanogen proliferation. Influent pH, ranging from 7.0 to 8.1, was also negatively correlated with all three methanogenic orders. The lower end of this range(around pH 7.0) aligned more closely with the optimal growth conditions for the methanogens identified, explaining the inverse relationship. OLR displayed a positive correlation with daily biogas production, but no significant relationship with methanogenic abundance. Similarly, influent COD concentration was uncorrelated with methanogen populations but positively associated with both DBP and COD removal efficiency, reflecting its role as an available substrate pool rather than a direct community structuring factor.

RDA analysis at the taxonomic order level provided finer resolution. HRT correlated positively with *Methanomicrobiales* (MMB) and negatively with *Methanobacteriales*(MBT), indicating that MMB dominated under longer HRT conditions (50 days), whereas MBT were better adapted to shorter HRTs (15 days). This finding aligns with previous reports that MMB species have higher hydrogen affinities than MBT, enabling them to thrive when hydrogen is scarce. *Methanosarcinales* (MSL) represented only 2.6% of themethanogen community and were thus considered ecologically minor in this system. Notably, HRT also showed a positive relationship with COD removal efficiency, linking operational conditions to both microbial structure and process outcomes. The RDA plots from this study visually captured these relationships, with environmental parameter vectors pointing toward the taxa and performance metrics they most strongly influenced. This case illustrates how RDA can be employed to disentangle complex interactions between operational parameters, microbial ecology, and system performance, enabling targeted operational adjustments to optimize both community stability and biogas productivity.

2.2.4 Principle Coordinate analysis

Principle Coordinate Analysis (PCoA), alsoknown as metric multidimensional scaling (mMDS), is a widely used ordination method in microbial ecology and other environmental sciences for exploring and visualizing similarities or dissimilarities among complex biological communities (Anderson, 2001; Legendre & Legendre, 2012). The technique begins with a matrix of pairwise distances between samples, which can be computed using a variety of distance or dissimilarity measures. In microbial community studies, these measures often include Bray–Curtis dissimilarity, Jaccard distance, or UniFrac distances when phylogenetic information is incorporated (Lozupone et al., 2005). The choice of metric is critical, as it determines how ecological relationships are represented in the ordination space (Faith et al., 1987).

Unlike Principal Component Analysis (PCA), which operates on raw variables and assumes linear relationships between them, PCoA operates directly on a dissimilarity matrix and can thus accommodate non-Euclidean distance measures. This makes it especially suitable for high-dimensional, compositional data such as those generated from next-generation sequencing (NGS) datasets, where the number of variables (taxa) often exceeds the number of samples by orders of magnitude, and the data often violate assumptions of normality (McMurdie & Holmes, 2013; Ramette, 2007).

The core idea of PCoA is to project samples into a lower-dimensional space—typically two or three dimensions—while preserving, as faithfully as possible, the pairwise distances from the original high-dimensional space. The axes in a PCoA plot are ordered according to the proportion of variation they explain, with the first axis capturing the largest share of variance, followed by the second, third, and so forth. The coordinates of each sample along these axes can then be interpreted in terms of their ecological similarity: samples positioned closer together in the plot are more similar in community composition, whereas those farther apart are more dissimilar (Ramette, 2007). In practice, PCoA isoften implemented as part of standard microbial community analysis pipelines (e.g., QIIME, mothur, vegan in R) (Caporaso et al., 2010; Oksanen, 2022; Schloss et al., 2009). The graphical output of PCoA is intuitively interpretable and provides a valuable first step in uncovering patterns in complex datasets before more formal statistical testing is applied. It should be noted, however, that the apparent clustering seen in a PCoA plot does not, by itself, imply statistical significance, formal tests such as PERMANOVA (permutational multivariate analysis of variance) are often used to confirm whether observed group separations are greater than would be expected by chance.

2.2.4.1 Application in sample comparison and clustering

One of the principal applications of PCoA in microbial ecology is the comparison of community structures among samples collected under different environmental conditions, treatments, or time points (Lozupone et al., 2005). By plotting samples in reduced-dimensional space, PCoA enables rapid visual assessment of whether communities from particular categories group together or are dispersed. For example, in an analysis of anaerobic digesters, samples from reactors operating under similar organic loading rates might cluster closely, while those from reactors with differing feed compositions or pH regimes might appear widely separated (Werner et al., 2014). Clustering patterns in a PCoA plot can reveal ecological gradients or discrete groupings within the dataset. These groupings often correspond to environmental or operational factors influencing microbial composition. For instance, in studies of river microbiomes, upstream and downstream samples may form distinct clusters, reflecting the influence of nutrient input, pollution, or hydrological variation. Similarly, in gut microbiome research, PCoA can distinguish between healthy individuals and those with disease states, with each group forming its own cluster in ordination space.

PCoA is also invaluable for temporal studies, where samples collected at successive time points can be connected to form trajectories in ordination space. Such trajectories can reveal community succession patterns, shifts following perturbations, or recovery dynamics after disturbances. In environmental biotechnology applications, temporal PCoA analysis can indicate whether microbial communities are stabilizing under a new operational regime or drifting toward an alternative state (Gundersen et al., 2021). Another important use of PCoA is in exploratory clustering. While hierarchical clustering methods can provide a dendrogram representation of similarity, PCoA complements this by offering a spatial, two- or three-dimensional visualization of relationships.

The combination of PCoA and clustering analyses can strengthen inferences: clusters identified statistically (e.g., via k-means or hierarchical methods) can be overlaid on a PCoA plot to assess their ecological coherence.

When integrating PCoA with environmental data, vector fitting orbiplot overlays can be used to map environmental gradients directly onto the ordination space. This allows researchers to identify which measured environmental variables are most strongly associated with the observed clustering patterns. For example, in wastewater treatment studies, vectors representing pH, ammonia concentration, or COD removal efficiency can be projected onto the PCoA plot, revealing whether these parameters align with separation between samples.

In summary, PCoA is a flexible and robusttool for visualizing microbial community similarities and differences. Its strength lies in its ability to condense complex, multidimensional data into an interpretable graphical format, making it an essential step in exploratory data analysis for microbial ecology, environmental monitoring, and biotechnology research. When combined with rigorous statistical testing and environmental metadata, PCoA becomes a powerful approach for uncovering the ecological drivers of microbial community structure and for guiding targeted experimental or operational interventions (Ramette, 2007).

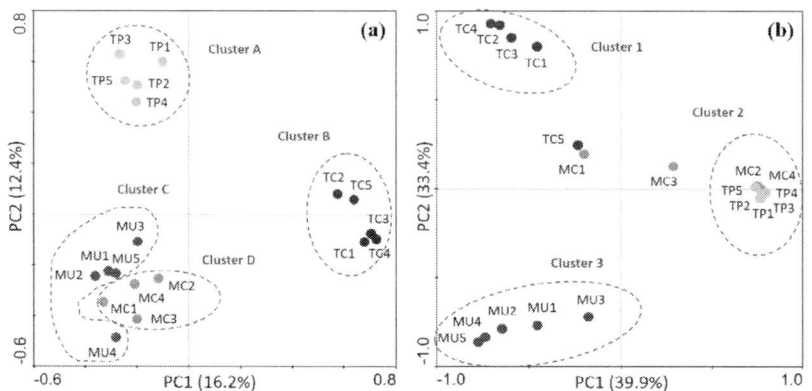

Fig 2.2 Principal coordinates analysis (PCoA) of (a) the bacteria and (b) the archaea of the digester samples. The PCoAplots were based on Sorensen (Bray-Curtis) distance and relative abundances of bacterial OTUs and archaeal OTUs were used as the dataset (Lee et al., 2017b).

A clear example of applying Principle Coordinate Analysis (PCoA) in anaerobic digestion research comes from a study investigating bacterial and archaeal communities in four full-scale thermophilic and mesophilic digesters treating food wastewater (Lee et al., 2017b). The goal was to visualize and interpret the similarity and dissimilarity of microbial community structures both within and between digesters, and to link these patterns to operational context. The analysis was based on the relative abundances of operational taxonomic units (OTUs) derived from sequencing data for 19 digester samples. Separate PCoA ordinations were produced for bacterial and archaeal communities, with clustering analysis applied to identify natural groupings of samples based on community composition.

For bacterial communities, cluster analysis at the 50% relative similarity criterion revealed four distinct clusters (A–D). Importantly, all samples grouped strictly according to their digester of origin. This indicated that variation between digesters far exceeded variation within a given digester. The conclusion was that each digester supported a distinct bacterial community structure, an observation consistent with earlier reports highlighting the site-specific nature of microbial consortia in full-scale AD systems. Further examination of shared taxa confirmed this uniqueness: only 2 of the 27 abundant genera (*Gelria* and *Proteiniphilum*) were found across all digesters, representing just 7% of the total. Moreover, some digesters harbored a high proportion of unique genera; for instance, the TC digester contained 75% unique abundant taxa, compared to 18–23% in the others. Patterns of genus sharing were also influenced by digester configuration: plug-flow systems (TP and MU) shared *Acholeplasma* and *Sphaerochaeta*, while CSTR systems (TC and MC) shared no abundant genera.

Archaeal community structures exhibited a different pattern. Cluster analysis at the 70% similarity threshold yielded three clusters, with each cluster largely corresponding to a specific digester or group of digesters: Cluster 1 comprised TC samples, Cluster 3 comprised MU samples, and Cluster 2 comprised TP samples together with two MC samples. However, significant within-digester archaeal shifts were observed in certain cases. For example, in the TC digester, the community shifted from *Methanothermobacter thermophilus* dominance to *Methanothermobacter beijingense* dominance. In the MC digester, a series of stepwise changes occurred: *Methanosaeta concilii* → *Methanoculleus bourgensis* → *Methanothermobacter beijingense* → *Methanoculleus bourgensis*. Remarkably, these substantial archaeal shifts occurred without any notable changes in COD removal efficiency, underscoring that process performance can remain stable even when community composition changes dramatically—a finding well documented in the AD literature.

The PCoA ordination plots for bacteria and archaea revealed that bacterial and archaeal communities were likely shaped by different sets of process parameters in the digestion of food wastewater. While bacterial communities appeared to be strongly digester-specific, archaeal communities showed greater within-digester variability and sensitivity to shifts in dominant methanogens. This case illustrates how PCoA, when combined with clustering analysis and taxonomic profiling, can provide both a visual and quantitative basis for interpreting microbial community relationships in anaerobic digestion systems. It allows researchers to identify whether differences are primarily driven by between-digester factors such as configuration and operating temperature, or by within-digester dynamics such as shifts in dominant functional groups. Importantly, the findings also highlight that microbial community changes do not always equate to performance instability, a nuance critical for process monitoring and management.

2.2.5 Non-metric multidimensional scaling

Non-metric Multidimensional Scaling (NMDS) is a widely employed ordination method in ecology, particularly suited for complex, high-dimensional datasets such as those generated in microbial community analyses. Unlike metric ordination techniques such as Principal Coordinate Analysis (PCoA), NMDS does not attempt to preserve actual distances between samples in a Euclidean sense. Instead, it focuses on maintaining the rank order of the distances or dissimilarities among samples. This distinction is critical: by working with ranks rather than raw distances, NMDS can handle data that deviate strongly from linearity, normality, or other assumptions required by parametric methods (Legendre & Legendre, 2012).

The foundation of NMDS lies in the construction of a dissimilarity (or distance) matrix among all pairs of samples. In microbial ecology, these dissimilarities are often computed using ecological distance measures such as Bray–Curtis, Jaccard, or Sørensen indices. NMDS then iteratively searches for a configuration of samples in a reduced-dimensional space (typically two or three dimensions) that best preserves the rank order of these dissimilarities. The quality of the ordination is assessed using a statistic known as stress, which quantifies the mismatch between the distances in the ordination space and the original rank order. Lower stress values indicate a better fit, with values below 0.1 generally considered excellent, between 0.1 and 0.2 acceptable, and above 0.3 indicating that the ordination is likely unreliable. Unlike eigenvalue-based techniques, NMDS is non-deterministic and often starts from a random initial configuration. As a result, different runs can yield slightly different solutions unless the same random seed is specified. For this reason, it is common practice to run NMDS multiple times from different starting configurations to ensure that the algorithm converges on a stable, low-stress solution (Bakker, 2024).

NMDS is especially valued for its flexibility in handling complex community datasets where many variables are zero for a given sample (the so-called "zero-inflated" nature of microbial OTU or ASV tables). Because it does not impose a linear relationship between dissimilarities and ordination distances, it is robust to non-linear species–environment relationships and heterogeneous variance structures, features that are typical in environmental microbiology datasets (Guseva et al., 2022).

2.2.5.1 Concept and application in microbial community analysis

In microbialecology, NMDS is frequently applied to visualize differences in community composition among samples collected under varying environmental conditions, treatments, or time points. By translating multivariate data into a two- or three-dimensional ordinationplot, NMDS allows researchers to identify patterns that might be obscured in raw abundance tables. The conceptual strength of NMDS lies in its rank-preserving nature. For example, if two microbial communities are more similar to each other than either is to a third community, NMDS will aim to preserve that ordering in the ordination space, even if the actual ecological distances cannot be perfectly represented.

This is especially important in microbial datasets, where dissimilarity measures may be stronglyinfluenced by rare taxa, compositional constraints, or uneven sequencing depths (Ramette, 2007).

In practical application, NMDS ordination can reveal grouping or separation of samples corresponding to environmental gradients, operational regimes, or biological states. For instance, in anaerobic digestion (AD) studies, NMDS can be used to compare bacterial and archaeal communities from digesters operating under different temperatures, organic loading rates, or feedstock types. Samples that plot close together in NMDS space can be interpreted as having similar community structures, whereas samples that plot far apart are more dissimilar. In one illustrative example from AD research, NMDS applied to bacterial community data might show distinct clustering by digester type, such as clear separation between thermophilic and mesophilic systems, while archaeal communities could instead group according to organic loading rate or ammonia concentration (Calusinska et al., 2018). Such differences in clustering patterns suggest that bacterial and archaeal communities respond to different sets of controlling parameters, a finding that can guide process optimization strategies.

Another application is in temporal monitoring of microbial communities. In longitudinal studies, NMDS can track changes in community structure over time. Samples collected at successive operational stages can be connected in the ordination space to form temporal trajectories, revealing patterns of succession, disturbance response, or recovery. In ADsystems, this approach can be used to assess whether a microbial community is stabilizing under steady-state operation or undergoing shifts that might precede performance changes (Goux et al., 2015). NMDS plots can also be augmented by overlaying environmental variables using vector fitting or surface fitting methods. This integration allows researchers to assess the correlation between microbial community patterns and measured process parameters, such as pH, volatile fatty acids (VFA), total ammonia nitrogen (TAN), or COD removal efficiency—directly within the ordination space. Such overlays often reveal key drivers of community variation, making NMDS not just a visualization tool, but also an entry point for hypothesis generation and targeted experimentation (Dai et al., 2016; Lee et al., 2008).

Non-metric multidimensional scaling (NMDS) has been successfully applied to explore temporal changes in microbial community composition within anaerobic digestion systems. In a comparative study of full-scale mesophilic and thermophilic digesters treating food waste-recycling wastewater, NMDS was used to visualize shifts in both bacterial and archaeal community structures over a series of sampling points (Kim et al., 2018). The analysis was performed on relative abundance data at the genus level, providing a detailed view of community composition dynamics.

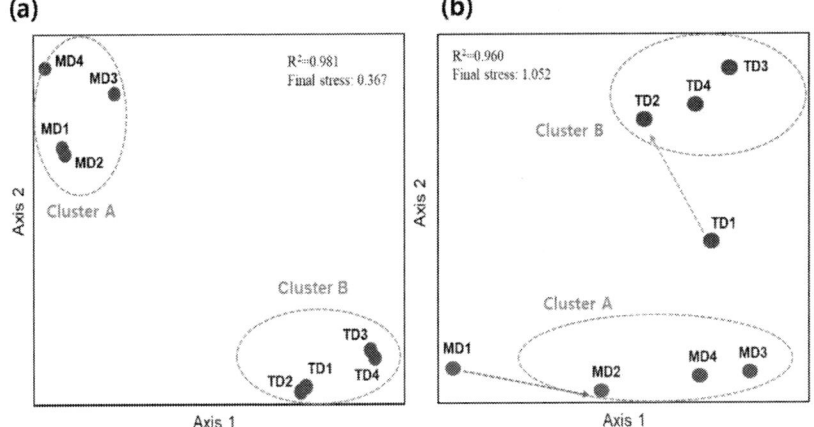

Fig 2.3 Non-metric multidimensional scaling of (a) bacterial communities and (b) archaeal communities in the digester samples. The NMDS plots were based on Sorensen (Bray-Curtis) distance and relative abundances of bacteria and archaea genera. Purple dotted linein (a): 75%: information remaining criterion (IRC); purple dotted line in (b) 85% IRC (Kim et al., 2018).

The NMDS ordinations were of high explanatory power. For the bacterial community, the cumulative R^2 was 0.98, with the first two dimensions capturing 60% of the cumulative variance. For the archaeal community, the cumulative R^2 was 0.96, with 87.0% of the variance explained. The bacterial ordination had a final stress value of 0.37 and final instability of 0.00006, while the archaeal ordination had a final stress of 1.05 with final instabilities of 0.00006 and 0.00001 for successive runs. These metrics indicated acceptable accuracy and stability of the NMDS configurations, consistent with published guidelines for interpreting NMDS results.

The ordination plots revealed strikingly different temporal patterns between bacteria and archaea. Bacterial community structures remained relatively stable over the sampling period in both the mesophilic (MD) and thermophilic (TD) digesters. Within each digester, bacterial samples clustered tightly, meeting a 75% relative similarity criterion, suggesting that bacterial populations were largely resilient to operational or environmental fluctuations during the study period. In contrast, archaeal communities exhibited significant temporal shifts. In the mesophilic digester, the first sampling point (MD1) was markedly distinct from subsequent samples (MD2, MD3, and MD4), which formed a separate cluster with 85%

relative similarity. This shift coincided with a change in the dominant methanogen genus—from *Methanobacterium* in MD1 to *Methanoculleus* in later samples. The transition was linked to changes in ammonium concentration; elevated ammonium levels and associated increases in volatile fatty acids (VFA) are known to favor syntrophic acetate oxidation coupled with hydrogenotrophic methanogenesis by *Methanoculleus*. However, ammonium concentrations above 3,000 mg/L can be toxic to many anaerobes, potentially limiting community stability under certain conditions.

A similar pattern was observed in the thermophilic digester. The first sample (TD1) was distinct from a later cluster (TD2, TD3, TD4). Here, the dominant archaeal genus shifted from *Methanoculleus* to *Methanothermobacter*. This transition was consistent with pH conditions during the sampling period. *Methanoculleus* thrives within a narrower pH range (6.2–7.8) and tends to decline when pH exceeds 8.0, whereas *Methanothermobacter* tolerates a broader range

(6.0–8.0) and can persist under slightly alkaline conditions. The thermophilic digester recorded pH values of 8.0–8.1 during TD2 and TD3, which likely suppressed *Methanoculleus* and allowed *Methanothermobacter* to increase in abundance. The study illustrates the capacity of NMDS to reveal nuanced temporal and environmental influences on microbial communities in anaerobic digesters. By differentiating between the relatively stable bacterial populations and the more dynamic archaeal populations, NMDS provided insights into the distinct ecological drivers governing each domain. Furthermore, the approach highlighted the importance of linking ordination patterns to process parameters, such as pH, ammonium concentration, and VFA levels, to interpret the functional significance of observed community shifts.

2.2.6 Canonical Correspondence Analysis

Canonical Correspondence Analysis (CCA) is a constrained ordination technique designed to directly relate species composition data to measured environmental variables. In contrast to unconstrained ordination methods such as Principal Coordinate Analysis (PCoA) or Non-metric Multidimensional Scaling (NMDS), which identify patterns in community structure without reference to external factors, CCA constrains the ordination so that the derived axes are linear combinations of environmental variables (Legendre & Legendre, 2012; Ramette, 2007). This makes CCA particularly powerful for testing specific hypotheses about the relationships between community composition and environmental gradients. CCA assumes unimodal species–environment relationships, meaning that each species exhibits a bell-shaped abundance curve along environmental gradients, with an optimal condition at which its abundance peaks and declining abundance on either side of that optimum. This unimodal response model is well suited to many microbial datasets, especially when sampling spans broad environmental ranges, such as varying pH levels, nutrient concentrations, or temperature regimes.

The analysis begins with two matrices: a species (or OTU/ASV) abundance matrix and an environmental variable matrix. Using an algorithm that maximizes the correlation between species composition and the environmental variables, CCA produces an ordination in which the positions of samples, species, and environmental vectors can be interpretedsimultaneously. Samples that are close to one another in the ordination space have similar community compositions and similar environmental characteristics. Species that plot near an environmental vector are positively associated with that parameter, while those plotting in the opposite direction are negatively associated. The length of each environmental vector indicates the strength of its correlation with community composition (Ter Braak & Verdonschot, 1995). An important advantage of CCA is that it quantifies the proportion of total variation in species composition that is explained by the measured environmental variables. This proportion, often expressed as the percentage of constrained variance, can be used to assess how well the measured parameters capture the main ecological gradients influencing community structure. Permutation tests are typically applied to determine the statistical significance of the overall model and of individual axes, providing a rigorous basis for interpretation (Ter Braak & Verdonschot, 1995).

2.2.6.1 Correlating microbial communities and performance parameters

In microbial ecology, CCA is widely applied when the research objective is to identify and visualize how specific environmental factors shape microbial community composition. For example, in anaerobic digestion studies, CCA can be used to relate methanogenic and bacterial community structures to operational parameters such as hydraulic retention time (HRT), organic loading rate (OLR), pH, total ammonia nitrogen (TAN),and volatile fatty acids (VFAs) (Kong et al., 2019). Because the axes of a CCA ordination are directly constrained by these variables, the resulting plot reveals not only how samples and taxa are distributed relative to each other, but also how they align with environmental gradients.

Consider an example where CCA is applied to archaeal community data from multiple digesters. The ordination might reveal that hydrogenotrophic methanogens such as Methanoculleus are positioned in the direction of high TAN and elevated VFAs, suggesting adaptation to ammonia-rich conditions often associated with syntrophic acetate oxidation. In contrast, acetoclastic methanogens like Methanosaeta may plot in the opposite direction, aligning with lower ammonia concentrations and more stable pH values (Kong et al., 2019). Such visualizations provide immediate ecological insights and can inform operational decisions, such as adjusting OLR or managing pH, to favor desired microbial functional groups. In addition to direct interpretation, CCA can be combined with statistical partitioning methods (e.g., variation partitioning) to separate the influence of different sets of variables, such as operational parameters versus feedstock characteristics, on community structure. This approach can reveal, for instance, whether differences between digesters are more strongly driven by feedstock type or by reactor operation (Li et al., 2015).

CCA is also valuable in temporal studies where the goal is to link microbial succession to measured environmental changes. For example, during the start-up phase of an anaerobic digester, CCA can show how initial fluctuations in pH, TAN, and VFAs correspond to shifts in dominant methanogenic orders, revealing the environmental drivers that govern community stabilization (Goux et al., 2015). Despite its strengths, CCA does have limitations. The assumption of unimodal species–environment responses means it is less suitable for datasets where relationships are strongly linear or where gradients are narrow. In such cases, redundancy analysis (RDA) may be more appropriate. Additionally, because CCA constrains ordination to measured variables, unmeasured but ecologically important gradients will not be represented in the constrained axes. For this reason, CCA is best used in conjunction with exploratory unconstrainedmethods to ensure that key ecological patterns are not overlooked.

In a study examining the effect of different microbial inocula on batch anaerobic digestion of fish waste, CCA was employed to directly relate the relative abundances of microbial genera tophysicochemical parameters including chemical oxygen demand (COD), lipids, proteins, total ammonia nitrogen (TAN), acetate, and propionate (Jannat et al., 2022). The ordination results demonstrated strong explanatory power: the first two CCA axes accounted for 70.3% of the variation in microbial abundance and the measured process variables. The analysis produced two well-defined clusters, as confirmed by ANOSIM ($R = 0.83$, $p < 0.05$). Cluster 1 (with >15% similarity) comprised the initial and endpoint samples from reactor R1 and the endpoint from R2, while Cluster 2 (with >50% similarity) contained all other initial and endpoint samples. These groupings reflected distinct microbial–environment relationships driven by differences in digestion stage and reactor conditions.

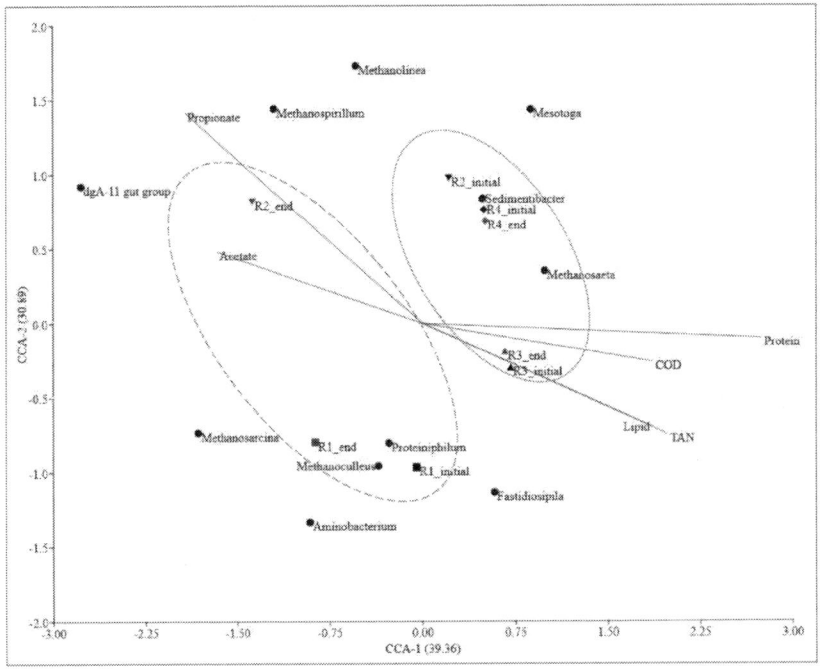

Fig 2.4 Canonical correspondence analysis (CCA) ordination diagram between the performance parameters and key genera obtained from 16S rRNA gene sequencing results. (initial: start point of the batch, end: endpoint of the batch)

Tripot analysis of the CCA ordination revealed clear genus–parameter linkages. Increases in acetate and propionate concentrations were positively correlated with higher relative abundances of *Proteiniphilum* and *Aminobacterium* during AD of fish waste. This relationship aligns with previous findings from chicken manure digestion, where syntrophic *Proteiniphilum* acted as an enhancer of volatile fatty acid (VFA) production, ultimately boosting methane yields. In contrast, *Fastidiosipila* abundance exhibited a negative correlation with acetate and propionate concentrations, a trend explained by its overall decline in relative abundance during digestion. Archaeal dynamics further highlighted the value of CCA in interpreting functional roles. In reactors R3 and R4, *Methanosaeta* emerged as the dominant methanogen, with a negative correlation to acetate and propionate levels. This pattern suggested active consumption of these VFAs by *Methanosaeta* for methane production. Interestingly, *Methanosaeta* abundance was positively correlated with TAN concentrations. This is consistent with previous studies in food waste digestion reporting *Methanosaeta* dominance (>90% relative abundance) at TAN levels of approximately 4.4 g/L and VFA concentrations of around 1.8 g/L.

The study's CCA biplot visually captured these relationships, with environmental vectors pointing toward taxa most strongly associated with specific process parameters. This direct visualization of microbial–environment linkages underscores howCCA can move beyond pattern detection to mechanistic interpretation. By integrating microbial composition with performance data, operators and researchers can identify microbial indicators linked to beneficial or inhibitory conditions and adjust process parameters accordingly to steer the microbial community toward improved stability and methane production.

Chapter 3. Kinetics of Microbial Growth and Substrate Utilization

3.1 General concepts of biokinetics in AD

Anaerobic digestion (AD) is a complex environmental bioprocess involving the concerted interaction between substrates, microbial communities, and the operational system under defined process conditions. At the core of optimizing and controlling these interactions lies biokinetics, thestudy of microbial growth, substrate utilization, and reaction rates (Adekunle et al., 2015; Leng et al., 2018).

In an AD system, the substrates serve as the primary source of carbon, energy, and nutrients for microorganisms. These substrates may include organic pollutants, nutrients, and other biodegradable compounds. The efficiency of AD relies heavily on how effectively these substrates are converted into methane, carbon dioxide, and other by-products by microbes, the biological agents driving the process. Bacteria, archaea, and other specialized microorganisms play distinct roles in degrading complex organics, syntrophically transferring intermediates, and ultimately facilitating methanogenesis (Venkiteshwaran et al., 2015).

The system itself comprising the physical and environmental conditions such as pH, temperature, mixing intensity, and reactor configuration, creates the operational environment in which microbial activities occur. These conditions directly influence microbial metabolism, substrate conversion rates, and process stability. Even slight deviations in operational parameters can lead to process inefficiencies, accumulation of intermediates (such as volatile fatty acids), or inhibition of methanogenic activity (Wainaina et al., 2019). Process control is the operational framework that ensures AD achieves its target objectives, whether that is maximum biogas production, specific pollutant removal, or nutrient recovery. This involves continuous monitoring of key parameters, applying optimization strategies, and making informed adjustments to maintain stability and efficiency (Alzate Ibañz et al., 2017).

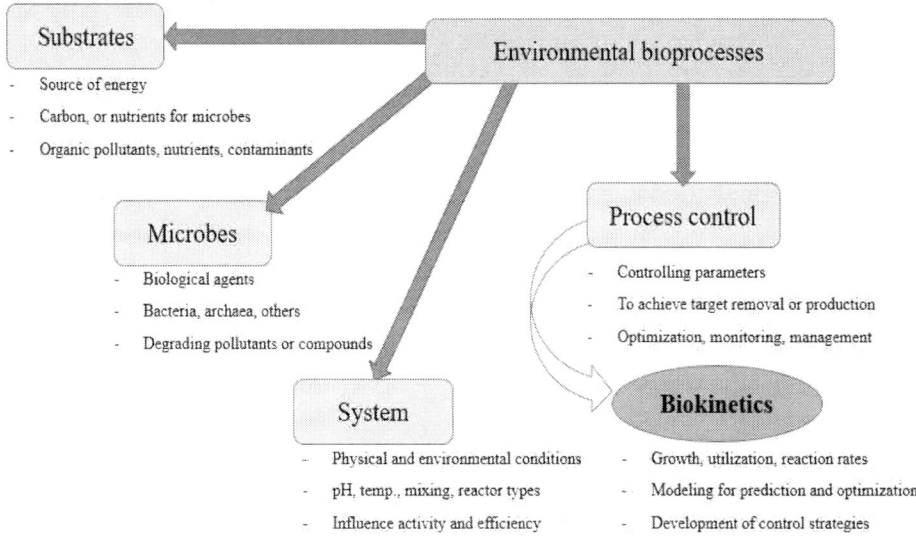

Fig 3.1 Biokinetics in environmental bioprocess

Here, biokinetics becomes indispensable. Understanding growth rates, substrate affinity, inhibition thresholds, and degradation kinetics allows operators and engineers to model system performance under different conditions. Such models are powerful tools for prediction and optimization, enabling proactive rather than reactive control. For example:

- Predictive modeling based on kinetic parameters can forecast system responses to changes in loading rate or temperature (Gerber & Span, 2008).
- Optimization strategies can be developed to maximize methane yield while minimizing energy input for mixing or heating (Lovato et al., 2021).
- Control strategies can be tailored to specific microbial bottlenecks, such as enhancing syntrophic activity or mitigating ammonia inhibition, by adjusting system parameters (Agyeman et al., 2021; Ziels et al., 2018).

In essence, biokinetics bridges the gap between biological potential and operational reality. By integrating biokinetic insights into AD process control, engineers can maintain stable performance, avoid process failures, and achieve higher efficiency in both energy recovery and waste treatment. Without a biokinetic framework, process control in

AD would be largely empirical and reactive, often resulting in trial-and-error adjustments. With it, the AD system becomes a predictable, optimizable, and sustainable environmental bioprocess (Venkiteshwaran et al., 2015).

3.1.1 Definition and importance

Biokinetics is the quantitative study of the rates of biological processes, particularly microbial growth and substrate utilization, and their relationship to environmental and operational factors. It serves as the link between microbiology and engineering design, allowing engineers to describe biological transformations mathematically and to predict the performance of bioreactors. In environmental and biochemical engineering, biokinetics plays a crucial role in:

- Design - determining reactor sizes, retention times, and loading rates.

- Operation - adjusting operating parameters for optimal performance.
- Control -anticipating process instability and applying corrective actions.
- Modeling and Simulation - predicting transient and steady-state behavior.

In AD, biokinetics is especially important because multiple microbial communities (hydrolytic bacteria, acidogens, acetogens, methanogens) work in sequence, each with its own growth characteristics. A good kinetic model helps balance these populations for stable methane production.

3.1.2 Microbial growth curve

Microorganisms, when introduced into a fresh growth environment, do not grow at a uniform rate over time. Instead, their population follows a characteristic pattern known as the microbial growth curve. This curve represents the change in biomass concentration or cell numbers as a function of time, and it reflects the physiological state of the microbial population as it adapts to its environment. The growth curve is most commonly plotted as:

- Linear scale: Biomass (X) vs. time (t) – useful for visualizing overall growth.
- Semi-logarithmic scale: log(X) vs. t – useful for quantifying growth rate.

A typical growth curve for a batch culture consists of four distinct phases (Fig 3.2):
  (1) Lag Phase
  (2) Exponential (Log) Phase
  (3) Stationary Phase
  (4) Death (Decline) Phase.

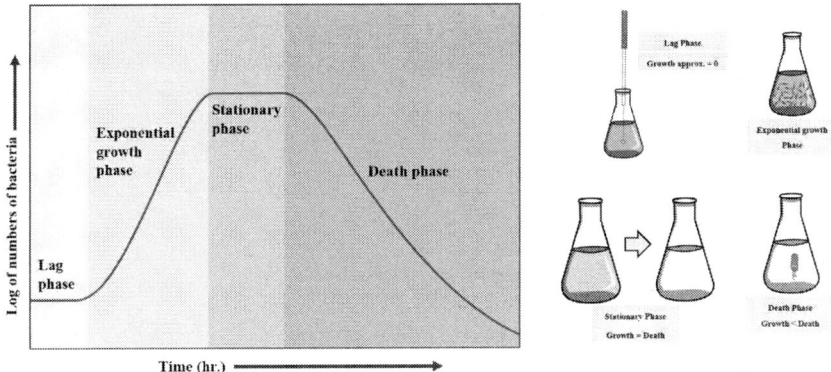

Fig 3.2 Microbial growth curve and phases

(1) Lag Phase

The period immediately after inoculation whenthere is no detectable increase in cell number or biomass. Microorganisms are metabolically active but are synthesizing enzymes, co-factors, and metabolic machinery needed for growth in the new medium. Key points of lag phase are:

- Duration depends on inoculum history, medium composition, and environmental factors.
- Cell mass may increase slightly due to water uptake and enzyme synthesis.

- No net increase in cell numbers, i.e., growth rate $\mu \approx 0$.

Mathematical representation of this phase: $dX/dt \approx 0$, $X(t) \approx X_0$

Where, X_0 = initial biomass concentration.

(2) Exponential (Log) Phase

Cells divide at a constant and maximum specific growth rate ($\mu \approx \mu_{max}$) because nutrients are abundant and inhibitory products are negligible. The rate of biomass growth is proportional to the biomass present:

$$\frac{dX}{dt} = \mu X \quad \quad (3.1)$$

For constant μ:

$$\frac{dX}{X} = \mu \, dt \quad \quad (3.2)$$

Integrating from t_0 to t:

$$\ln X - \ln X_0 = \mu \, (t - t_0) \quad \quad (3.3)$$

Or:

$$X(t) = X_0 e^{\mu(t-t_0)} \quad \quad (3.4)$$

Where. X(t) = biomass at time t; X_0 = biomass at start of exponential phase; μ = specific growth rate (time^{-1})

Doubling time (t_d) is defined as the time required for the biomass to be doubled, i.e., X(t_d) = 2X_0. Substituting into exponential equation 3.4,

$$2X_0 = X_0 e^{\mu t_d}$$

$$t_d = \frac{\ln 2}{\mu} \quad \quad (3.5)$$

Thus, for a given μ, doubling time is constant in the exponential phase.

(3) Stationary Phase

In this phase, growth rate slows to zero due to nutrient depletion, oxygen limitation, or accumulation of inhibitory products. Cell death balances cell division, resulting in constant biomass concentration. Mathematical representation of this phase is, $dX/dt \approx 0$, $X(t) \approx X_{max}$. However, metabolic activity may continue, with cells producing secondary metabolites.

(4) Death Phase

In this phase, cell death exceeds cell division due to prolonged nutrient limitation or toxicity. Viable cell numbers decline exponentially. Assuming first-order decay:

$$\frac{dX}{dt} = -bX \quad \quad \quad (3.6)$$

Integrating:

$$X(t) = X_{max}\, e^{-b(t-t_s)} \quad \quad \quad (3.7)$$

Where, b = decay constant (time-1); t_s = time at start of death phase.

3.2 Importance of biokinetics in AD process control

AD is a complex biochemical and microbial process that transforms organic matter into methane-rich biogas and digestate under oxygen-free conditions. The process comprises multiple stages, hydrolysis, acidogenesis, acetogenesis, and methanogenesis, each governed by distinct microbial communities operating under specific kinetic principles (Gaby et al., 2017). For process engineers and operators, maintaining stability, maximizing performance, and avoiding failures require a deep understanding of the kinetic behaviors of these microbial groups. Biokinetics, defined as the study of the rates atwhich biological processes occur, is not merely an academic or modeling construct. It forms the core of dynamic process control in anaerobic digestion systems. Without a grasp of how microbial activity responds to changing substrates, environmental conditions, and operating regimes, any control strategy remains reactive rather than predictive (Maleki et al., 2018). This section explores the vital importance of biokinetics in achieving intelligent, adaptive, and optimized control of AD systems.

Process control in anaerobic digestion refers to the regulation of operational parameters, such as feed rate, temperature, mixing intensity, pH, alkalinity, and retention times, to ensure optimal performance. At its core, process control is about managing the behavior of microbial populations that drive biochemical conversions. Unlike purely physical or chemical systems, biological systems such as AD are inherently nonlinear, time-variant, and sensitive to external perturbations. Therefore, effective control strategies must account for the biokinetics of microbial growth, substrate degradation, and product formation to anticipate system responses and prevent upset conditions (Alzate Ibañz et al., 2017).

One of the most crucial aspects of process control is the real-time monitoring of process variables and interpreting them through the lens of biokinetics. Some commonly monitored parameters include:

- Volatile fatty acids (VFAs): Accumulation indicates acidogenic rates exceeding methanogenic rates, a kinetic imbalance.
- Biogas production rate and composition: A decline in methane yield may signal substrate depletion or methanogen inhibition.
- Alkalinity and pH: Changes reflect acid production and buffering capacity, tightly linked to microbial activity.
- Ammonia and hydrogen sulfide concentrations: Indicators of substrate composition and potential inhibitors affecting microbial kinetics (Ghofrani-Isfahani et al., 2020).

These data points are often used in conjunction with kinetic models to estimate microbial activity levels, growth rates, and potential inhibitions, forming the basis of predictive control algorithms (Ghofrani-Isfahani et al., 2020).

One of the primary control levers in AD systems is the organic loading rate (OLR), the amount of substrate fed to the system over time. Biokinetics helps answer the fundamental control questions: How much substrate can be safely fed without overloading the system? What is the critical loading threshold beyond which VFAs will accumulate? (Rincó et al., 2008). How will different substrate compositions affect the hydrolysis and methanogenesis rates? Using biokinetic models like Monod kinetics or first-order hydrolysis, operators can estimate the maximum admissible OLR and design feed-forward controls. For example, when feeding protein-rich waste, which leads to ammonia formation, operators may reduce the feed rate or adjust pH in anticipation of kinetic inhibition. Moreover, in co-digestion scenarios, kinetic coefficients such as hydrolysis rates and specific methanogenic activity (SMA) are used to determine optimal mixing ratios and avoid competitive inhibition or substrate suppression.

Process stability in anaerobic digestion is largely governed by the dynamic equilibrium between different microbial populations. Since each group (acidogens, acetogens, methanogens) operates at different kinetic rates, imbalances can quickly lead to process failure. Faster-growing acidogens may rapidly generate VFAs, outpacing the capacity of slower-growing methanogens, leading to acidification (Jia et al., 2023). Ammonia inhibition slows down methanogenesis, causing a kinetic bottleneck. Hydrogen partial pressure influences the thermodynamic feasibility of acetogenesis, requiring kinetic synchronization. Biokinetic monitoring allows for the establishment of early warning systems that can detect when one microbial group is dominating or lagging behind, long before catastrophic failures (e.g., souring, gas collapse) occur (Wu et al., 2019). Operators can then take remedial action, such as adjusting temperature, feed composition, or retention times.

Optimization of AD processes, whether for maximizing methane yield, minimizing retention time, or enhancing nutrient recovery, relies heavily on kinetic understanding: Batch digesters are optimized for reaction time based on hydrolysis and methanogenic rates. Continuous stirred-tank reactors (CSTRs) require careful balancing of SRT and OLR, both of which are dictated by microbial kinetics. Two-stage digesters benefit from separating kinetic phases (e.g., fast hydrolysis vs. slow methanogenesis), leading to improved stability and performance (Pramanik et al., 2019). Operators can perform kinetic parameter estimation using lab-scale experiments and use these values to simulate full-scale reactor behavior under different scenarios.

Emerging digital technologies are redefining process control through the use of digital twins—real-time simulations of AD plants that continuously update using sensor data. At the heart of these digital twins are kinetic models calibrated with live data. By integrating SCADA systems, online monitoring instruments (e.g., gas flowmeters, VFA sensors), and kinetic models, digital twins can: Forecast methane production under different feed regimes. Predict the impact of operational changes or disturbances. Recommend control actions to maintain stability and efficiency. The kinetic backbone of these systems makes them capable of learning and adapting, ushering in a new era of intelligent anaerobic digestion (Schroer et al., 2023).

3.3 Fundamental kinetic models in AD

Understanding and modeling microbial kinetics is essential to predicting, controlling, and optimizing anaerobic digestion (AD) processes. Kinetic models describe the relationship between the rate of biological reactions and key environmental or operational parameters such as substrate concentration, microbial biomass, and retention time. These models form the foundation of biochemical reactor design, scale-up, process simulation, and control. Among the array of kinetic formulations available, three fundamental models have found widespread application in anaerobic digestion research and practice: The Monod, HaldaneContois, and First-order models. Each model offers unique advantages and limitations, depending on the microbial pathway being modeled, the nature of the substrate, and the scale of operation.

3.3.1 Monod Model

The Monod model, proposed by Jacques Monod in 1942, is arguably the most widely used kinetic model in biological wastewater treatment. It expresses microbial growth rate as a function of limiting substrate concentration, drawing analogies from enzyme kinetics (Jannasch & Egli, 1993). The Monod equation is given by:

$$\mu = \frac{\mu_{max} \cdot S}{K_s + S} \quad\quad\quad (3.8)$$

Where:

μ = specific growth rate of microorganisms (day^{-1})

μ_{max} = maximum specific growth rate (day^{-1})

S = concentration of limiting substrate (g/L)

K_S = half-saturation constant (g/L), the substrate concentration at which $\mu = \mu_{max} / 2$

The Monod model is extensively used to describe the growth kinetics of various microbial groups, including: Hydrolytic bacteria (e.g., degrading complex polymers) Acidogens and acetogens (e.g., converting soluble organics into VFAs) Methanogens (e.g., converting acetate, H_2/CO_2 into methane). For instance, methanogenic activity from acetate can be modeled using Monod kinetics, where the substrate is acetate and the growth rate reflects the ability of acetoclastic methanogens to metabolize it. Assumptions and Limitations Assumes a single limiting substrate. Neglectsmicrobial competition, inhibition, and syntrophic. Assumes a well-mixed system without diffusion limitations. Parameters (μ_{max}, K_S) must be determined empirically (KováováKovar et al., 1998). Despite these limitations, the Monod model remains a foundational tool in AD modeling due to its conceptual simplicity and interpretability. It forms the basis of advanced models like ADM1 (Anaerobic Digestion Model No.1) (Batstone et al., 2002).

3.3.2 Haldane Model

While the Monod model successfully describes microbial growth under moderate substrate concentrations, it fails when substrate levels become excessively high and start to inhibit microbial activity. Such inhibition is common in environmental and industrial processes, including anaerobic digestion, where high concentrations of ammonia, volatile fatty acids (VFAs), or certain industrial organics can slow or stop microbial growth. To account for this, the Haldane model (also known as the Andrews model) extends Monod's concept by adding a substrate inhibition term in the denominator.

The Haldane equation for specific growth rate is:

$$\mu = \frac{\mu_{max} \cdot S}{K_s + S + \frac{S^2}{K_I}} \quad \quad (3.9)$$

Where:

K_I = Substrate inhibition constant

A large value of K_i implies that substrate is less inhibitive to microorganisms. However, a low value of K_i shows a strong inhibition effect to microorganisms and in this case the substrate needs to be diluted to increase the growth of microbial community and increase the treatment efficiency. Limitations of this model are: (i) Assumes inhibition is only due to the primary substrate; does not handle multiple inhibitors; (ii) Requires experimental determination of K_i, which can vary significantly between systems and conditions; (iii) Assumes well-mixed conditions without mass transfer limitations.

3.3.3 Contois Model

The Contois model was developed for systems where microbial growth is significantly influenced by the ratio of substrate concentration to biomass concentration. This is common in high-cell-density environments (e.g., activated sludge systems, anaerobic digesters with high solids) where, substrate availability per cell is limited and microbial competition for substrate is intense.

The Contois equation is:

$$\mu = \frac{\mu_{max} \cdot S}{K_c X + S} \quad\quad\quad (3.10)$$

Where:

K_c = Contois saturation constant (dimensionless with respect to S/X)

The Contois equation has the similar structure with the Monod equation mathematically. The difference lies in the denominator (i.e., K_S+S in Monod equation and K_c $X+S$ in Contois equation). Mathematically K_S in Monod is equal to K_c X in Contois (i.e., $K_S = K_c$ X). However, in Contois the substrate affinity of microorganism (K_c X) is not a constant, but a linear function of microbial concentrations.

Although Monod and Contois equations have similar mathematical expressions, they have a huge difference in physiological meaning. In Contois equation, a small K_c will lead to small $K_c X$ value (or strong affinity of microorganism to substrate) at low concentration of microorganism and the specific growth rate will be equal to the maximum specific growth rate, $\mu = \frac{\mu_m S}{K_c X + S} = \mu_m$. As X increases the $K_c X$ value will increase linearly and the specific growth rate becomes the first order equation, $\mu = \frac{\mu_m S}{K_c X + S} = \frac{\mu_m S}{K_c X}$; Therefore, the specific growth rate at high microbial concentrations is lower than that at high concentrations. As a result, the specific growth rate of microorganism is a function of both substrate and microorganism concentrations in Contois equation. However, in Monod, this rate is only a function of substrate which means that the Monod equation is a simplified version of Contois equation.

3.3.4 Lotka-Volterra Model

The Lotka-Volterra model is a classical mathematical framework originally developed in the

1920s by Alfred J. Lotka (1925) and Vito Volterra (1926) to describe predator–prey dynamics in ecological systems. Although it was first applied to macroscopic ecosystems (e.g., wolves and rabbits), it is equally applicable to microbial consortia, where one population's growth depends on the presence or consumption of another. In AD and other bioprocesses, the Lotka-Volterra framework can describe:

- Predator-prey-like relationships between bacteriophages and bacteria (Crowley et al., 1980).
- Syntrophic interactions, e.g., hydrogen-producing acetogens (prey) and hydrogenotrophic methanogens (predators for H_2).
- Competition for a common substrate.
- Mutualism, where two species benefit and depend on each other.

The simplest Lotka–Volterra predator–prey model consists of two coupled differential equations:

$$\frac{dX}{dt} = r\,X - a\,X\,Y \qquad (3.11)$$

$$\frac{dY}{dt} = -m + b\,X\,Y \quad\quad\quad\quad\quad\quad (3.12)$$

Where, X = Prey population (cell number); Y = Predator population (cell number); r = Intrinsic growth rate of prey (time^{-1}); a = Predator rate coefficient (volume/mass.time); m = Mortality rate of predator (time^{-1}); b = Biomass yield of predator from prey consumed (mass/mass).

In environmental microbiology, "predator" and "prey" are broad terms:
- Predator: Virus, protozoa, predatory bacteria (e.g., Bdellovibrio), or a microbial guild consuming a metabolite produced by another guild.
- Prey: Bacteria, archaea, or substrate-providing population.

In anaerobic digestion:

- Syntrophic acetate-oxidizing bacteria (SAOB) produce H_2, which is immediately consumed by hydrogenotrophic methanogens.
- Although not classical predation, the dynamics can be mathematically similar; methanogens "control" H_2 levels, indirectly influencing SAOB growth rate.

The multispecies Lotka-Volterra model has the following forms:

$$\frac{dX_i}{dt} = \beta_i X_i = \frac{\mu_{mi} S}{K_{Si} + S} X_i \quad\ldots\ldots\ldots\ldots\ldots\ldots\ldots\ldots\ldots\ldots \quad (3.13)$$

$$\frac{dX_i}{dt} = \left(\beta_i + \sum_{j=1}^{n} \alpha_{ij} X_j\right) X_i \quad\ldots\ldots\ldots\ldots\ldots\ldots\ldots\ldots\ldots \quad (3.14)$$

Where, β_i represents the intrinsic growth rate of the species i and α_{ij} the influence of the species j on the growth rate of species i. This influence is positive or negative or independent according to the sign of $_{ij}$. The intrinsic growth rate $_i$ is either the Monod, Haldane or Contois equation:

$$\beta_i = \frac{\mu_{mi} S}{K_{Si} + S} \text{ or } \beta_i = \frac{\mu_{mi} S}{K_{Si} + S + \frac{S^2}{K_{I,i}}} \text{ or } \beta_i = \frac{\mu_{mi} S}{K_{Xi} X_i + S} \quad \ldots\ldots\ldots\ldots \quad (3.15)$$

The expansion of Lotka-Volterra model for two-species with Monod equation as follow:

$$\frac{dX_i}{dt} = \left(\beta_i + \sum_{j=1}^{n} \alpha_{ij} X_j\right) X_i \quad \ldots\ldots\ldots\ldots\ldots\ldots\ldots\ldots\ldots \quad (3.16)$$

The general equation of Lotka-Volterra is expanded for the case of two species as follows:

$$\frac{dX_1}{dt} = \left(\frac{\mu_{m1} S}{K_{S1} + S} + \alpha_{21} X_2\right) X_1 \quad \ldots\ldots\ldots\ldots\ldots\ldots\ldots\ldots \quad (3.17)$$

$$\frac{dX_2}{dt} = \left(\frac{\mu_{m2} S}{K_{S2} + S} + \alpha_{12} X_1\right) X_2 \quad \ldots\ldots\ldots\ldots\ldots\ldots\ldots\ldots \quad (3.18)$$

In these equations, α_{21} is the interaction term in which X_2 exerts on X_1 and vice versa. As mentioned above, a positive α_{12} is it means that the existence of X_2 will stimulate the growth of X_1. When α_{12} is zero, they grow independently. On the other hand, a negative tells us that X_2 hampers the growth of X_1. From the compacted form of the Lotka-Volterra model, the model for three, four or more microbe cases can be expanded.

It is noted that the interaction term is represented by a (certain) value of interaction parameter, α, but the effect of one microbial (e.g. X_2) to the other (e.g. X_1) is a linear function and this effect will increase proportionally to the concentration of (e.g. X_2). Because of the linearly function effect, the interaction effect become very large at high concentration of X_2. On the other hand, this effect become negligible as the concentration of X_2 is small. If we consider the overall ordinary differential equation, the interaction term actually increases one nonlinear order (i.e., from second order to third order system). Consequently, the numerical solution of this third-order equation (Eq. 3.14) is getting more sophisticated compared to that of the second-order equation (Eq. 3.13)

3.3.4.1 Case studies: Interaction modeling

A compelling case of interaction modeling within anaerobic microbial systems is provided by a study that explores how *Clostridium cadaveris* and *Clostridium sporogenes* interact during acidogenic fermentation of peptone as the sole carbon source. The researchers developed species-specific qPCR primer–probe sets, enabling precise quantification of each species' 16S rRNA gene in both pure and mixed culture settings. In pure cultures, they determined key biokinetic parameters, including maximum specific growth rate (μ_{max}), half-saturation constant (K_s), yield (Y), and decay coefficient (K_d), allowing for a detailed characterization of each species' growth kinetics (Koo et al., 2019).

Moving beyond pure cultures, the study modeled how the two species interact during co-cultivation. Interestingly, while mixed cultures exhibited accelerated substrate consumption, individual microbial growth rates declined, suggesting competitive or inhibitory interspecies interactions. To capture these dynamics, the authors proposed a new interspecific interaction model that balances biological realism with analytical simplicity and predictive accuracy (Jannat et al., 2024). In further modeling efforts that emulated continuous reactor conditions under varying hydraulic retention times (HRTs), the team used Runge–Kutta numerical methods coupled with non-linear regression to fit the differential equations governing biomass and substrate dynamics. This allowed them to simulate both transient and steady-state behaviors. Strikingly, while *C. sporogenes* maintained population stability across a range of HRTs, *C. cadaveris* declined markedly under short retention times, with washout predicted at HRTs below approximately 1.9 hours, highlighting its comparatively lower resilience in continuous systems (Koo et al., 2020). Overall, this study exemplifies how integrating experimental microbiology with mathematical modeling can yield deep insight into interspecies relationships in anaerobic systems. The ability to dissect species-specific growth characteristics, model interaction effects on substrate turnover, and predict washout thresholds in continuous operations makes this a powerful methodological blueprint. Such a modeling framework is highly adaptable to more complex mixed microbial systems, ranging from anaerobic digesters to waste treatment bioreactors, where understanding behind-the-scenes competition and cooperation is key to optimizing performance and stability.

3.4 Fundamental biokinetic equations

These are the building blocks for mass balances.

1. Microbial growth:

$$\frac{dX}{dt} = \mu X \quad \quad (3.19)$$

The equation indicates that the rate of biomass increase (dX/dt) is directly proportional to the biomass concentration (*X*) and the specific growth rate (μ).

2. Microbial yield:

$$\frac{dX}{dS} = Y \quad \quad (3.20)$$

It describes the relationship between the rates of change of biomass concentration *(X)*and substrate concentration *(S)*. The yield coefficient *(Y)* is determined experimentally and represents the efficiency of biomass production from substrate utilization.

3. Substrate utilization:

$$\frac{dS}{dt} = -\frac{1}{Y}\mu X \quad \dots\dots\dots\dots\dots\dots\dots\dots\dots\dots\dots\dots \quad (3.21)$$

It is derived from the mass balance equation for the substrate. The negative sign indicates that substrate concentration decreases over time due to microbial consumption.

4. Microbial decay:

$$\frac{dS}{dt} = -bX \quad \dots\dots\dots\dots\dots\dots\dots\dots\dots\dots\dots\dots \quad (3.22)$$

Microbial decay kinetics describe the rate at which biomass decreases over time due to cell death or inactivation.

Table 3.1 Common biokinetic parameters

Symbol	Units	Description
X	g/L	
S	g/L	
μ	1/d	
μ_m	1/d	
K_s	g/L	
K_I	g/L	
Y	g/g	
b	1/d	

3.4.1 Mass balance equation development

General Form:

Accumulation = Input − Output + Generation − Consumption

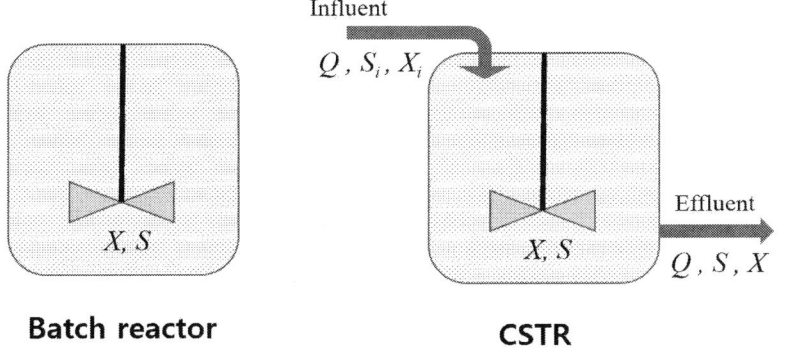

Fig 3.3 Flow diagram of batch and CSTR reactors

Here,

Q = Flow rate (L/d);

V = Volume (L);

S_i = Influent substrate concentration (g/L);

S = Substrate concentration in reactor (g/L);

X_i = Biomass concentration in influent (g/L);

X = Biomass concentration in reactor (g/L).

For batch reactor case, no flow terms.
- Biomass

$$\frac{dX}{dt} = \mu X - bX \quad\quad\quad (3.23)$$

- Substrate

$$\frac{dS}{dt} = -\frac{1}{Y}\mu X \quad\quad\quad (3.24)$$

For continuous stirred tank reactor (CSTR) case, dilution rate D = Q/V.

- Biomass

$$\frac{dX}{dt} = D(X_i - X) + \mu X - bX \quad \quad (3.25)$$

- Substrate

$$\frac{dS}{dt} = D(S_i - S) - \frac{1}{Y}\mu X \quad \quad (3.26)$$

3.4.2 Analytical solutions

3.4.2.1 Solutions for batch reactors

General mass balance equation for substrate:

Net change = Substrate utilization

$$\left(\frac{dS}{dt}\right)_{net} = \left(\frac{dS}{dt}\right)_u = -\frac{1}{Y}\left(\frac{dX}{dt}\right)_g \quad \quad (3.27)$$

$$\left(\frac{dS}{dt}\right)_u = \left(-\frac{\mu_{max} X}{Y}\right)\frac{S}{K_s + S} \quad \quad (3.28)$$

$$\int_{S_0}^{S} \frac{K_s + S}{S} dS = -\frac{\mu_{max} X}{Y} \int_0^t dt$$

$$K_s \ln\frac{S}{S_0} + S - S_0 = -\frac{\mu_{max} X}{Y} t$$

$$K_s \ln\frac{S_0}{S} + S_0 - S = \frac{\mu_{max} X}{Y} t$$

$$\frac{\ln\frac{S_0}{S}}{S_0 - S} = \frac{\mu_{max}}{YK_s} \frac{X}{S_0 - S} t - \frac{1}{K_s}$$

Linear solution with slope = $\frac{\mu_{max}}{YK_s}$ and

Y-intercept = $-\frac{1}{K_s}$

General mass balance equation for biomass:

Net change = Growth − Decay

$$\left(\frac{dX}{dt}\right)_{net} = (\mu - k_d)X \quad \ldots \ldots \ldots \ldots \ldots \ldots \ldots \ldots \ldots \ldots (3.29)$$

$$\int_{X_o}^{X} \frac{dX}{X} dS = (\mu - k_d) \int_0^t dt$$

$$\ln \frac{X}{X_o} = (\mu - k_d) t$$

$$X = X_0 \, e^{(\mu - k_d) t}$$

3.4.2.2 Solutions for CSTRs

If X_i = 0 for a CSTR system, general mass balance equation for biomass:

Net change = Input − Output + Growth − Decay

$$\left(\frac{dX}{dt}\right)_{net} V = 0 - QX + (\mu - k_d) XV$$

At steady state, $(dX/dt)_{net}$ = 0 & divided by XV,

$$\mu - k_d = \frac{QX}{VX} = \frac{Q}{V} = \frac{1}{\tau} \; ; \; \text{Where } \tau = HRT$$

$$\mu = \frac{1}{\tau} + k_d = \frac{1 + k_d \tau}{\tau} = \frac{\mu_{max} \cdot S}{K_s + S}$$

$$S = \frac{K_s(1 + k_d \tau)}{(\mu - k_d)\tau - 1} \quad \dots\dots\dots\dots\dots\dots\dots\dots\dots\dots \quad (3.30)$$

General mass balance equation for substrate:

Net change = Input − Output − Utilization

$$\left(\frac{dS}{dt}\right)_{net} = \frac{Q}{V}(S_i - S) - \left(\frac{dS}{dt}\right)_u$$

At steady state, $(dS/dt)_{net} = 0$,

$$\left(\frac{dS}{dt}\right)_u = \frac{Q}{V}(S_i - S) = -\frac{1}{Y}\left(\frac{dX}{dt}\right)_g = -\frac{1}{Y}\mu X$$

$$\frac{Q}{V} = \frac{1}{\tau} \text{ and } \mu = \frac{1}{\tau} + k_d$$

$$X = \frac{Q}{V}\frac{Y(S_i - S)}{\mu} = \frac{Y(S_i - S)}{1 + k_d \tau} \quad \dots\dots\dots\dots\dots\dots\dots\dots \quad (3.31)$$

3.4.3 Numerical solutions
3.4.3.1 4th order Runge-Kutta Method

When analytical solutions to ordinary differential equations (ODEs) are not readily obtainable, numerical integration methods provide a practical alternative. Among these, the Fourth-Order Runge–Kutta (RK4) method stands out as one of the most widely used techniques for its balance of computational efficiency, accuracy, and conceptual simplicity (Musa et al., 2010). RK4 is particularly useful in engineering and biological modeling, where mass balance equations describe the dynamic evolution of system variables over time (Elpianora et al., 2024). In essence, the RK4 method advances the solution of a differential equation from a known value at time t_i to the next time point t_{i+1} = = t_i+hby evaluating the rate of change (slope) of the function at strategically chosen points within the time interval h. This approach combines information from the beginning, middle, and end of the interval to approximate the true trajectory of the system more accurately than simpler methods such as Euler's method. The step-by step equations of RK4 method are following:

$$k_1 = f(t_i, S_t, X_t)$$

$$k_2 = f(t_i + \frac{h}{2}, S_t + \frac{h}{2}K_1, X_t + \frac{h}{2}K_1)$$

$$k_3 = f(t_i + \frac{h}{2}, S_t + \frac{h}{2}K_2, X_t + \frac{h}{2}K_2)$$

$$k_4 = f(t_i + h, S_t + hK_3, X_t + h K_3)$$

$$S_{t+1} = S_t + \frac{h}{6}[K_1 + 2K_2 + 2K_3 + K_4]$$

$$X_{t+1} = X_t + \frac{h}{6}[K'_1 + 2K'_2 + 2K'_3 + K'_4]$$

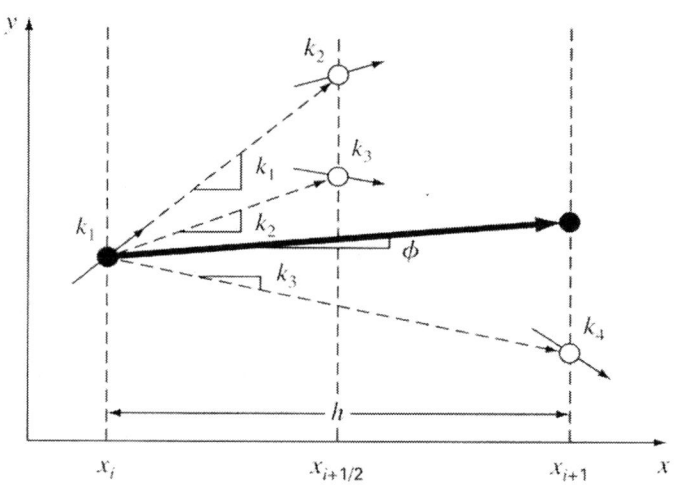

Fig 3.4 Graphical explanation of RK4 method

3.4.3.2 Solution for batch reactors

For substrate:

$$\frac{dS}{dt} = -\frac{\mu_m}{Y}\frac{SX}{K_s+S} \quad \dots\dots\dots\dots\dots\dots\dots\dots\dots\dots\dots (3.32)$$

Applying RK4 formulation:

$$K_1 = -\frac{\mu_m}{Y}\frac{S_t X_t}{K_s+S_t}$$

$$K_2 = -\frac{\mu_m}{Y}\left(\frac{\left(S_t+\frac{\Delta t}{2}S_1\right)\left(X_t+\frac{\Delta t}{2}X_1\right)}{K_s+\left(S_t+\frac{\Delta t}{2}S_1\right)}\right)$$

$$K_3 = -\frac{\mu_m}{Y}\left(\frac{\left(S_t+\frac{\Delta t}{2}S_2\right)\left(X_t+\frac{\Delta t}{2}X_2\right)}{K_s+\left(S_t+\frac{\Delta t}{2}S_2\right)}\right)$$

$$K_3 = -\frac{\mu_m}{Y}\left(\frac{(S_t+\Delta t\, S_3)(X_t+\Delta t\, X_3)}{K_s+(S_t+\Delta t\, S_3)}\right)$$

$$S_{t+1} = S_t + \frac{\Delta t}{6}\left[K_1 + 2K_2 + 2K_3 + K_4\right]$$

For biomass:

$$\frac{dX}{dt} = \left(\frac{\mu_m \cdot S}{K_s + S} - k_d\right) X \quad \ldots\ldots\ldots\ldots\ldots\ldots\ldots\ldots\ldots (3.33)$$

Applying RK4 formulation:

$$K'_1 = \left(\frac{\mu_m \cdot S_t}{K_s + S_t} - k_d\right) X_t$$

$$K'_2 = \left(\frac{\mu_m \left(S_t + \frac{\Delta t}{2} S_1\right)}{K_s + \left(S_t + \frac{\Delta t}{2} S_1\right)} - K_d\right)\left(X_t + \frac{\Delta t}{2} X_1\right)$$

$$K'_3 = \left(\frac{\mu_m \left(S_t + \frac{\Delta t}{2} S_2\right)}{K_s + \left(S_t + \frac{\Delta t}{2} S_2\right)} - K_d\right)\left(X_t + \frac{\Delta t}{2} X_2\right)$$

$$K'_4 = \left(\frac{\mu_m \left(S_t + \Delta t\, S_3\right)}{K_s + \left(S_t + \Delta t\, S_3\right)} - K_d\right)\left(X_t + \Delta t\, X_3\right)$$

$$X_{t+1} = X_t + \frac{\Delta t}{6}\left[K'_1 + 2 K'_2 + 2 K'_3 + K'_4\right]$$

3.4.3.3 Solution for CSTRs

For substrate:

$$\frac{dS}{dt} = \frac{S_i - S}{\tau} - \frac{\mu_m}{Y}\left(\frac{SX}{K_s + S}\right) \quad \ldots\ldots\ldots\ldots\ldots\ldots\ldots\ldots\ldots\ldots (3.34)$$

Applying RK4 formulation:

$$K_1 = \frac{S_i - S_t}{\tau} - \frac{\mu_m}{Y}\left(\frac{S_t X_t}{K_s + S_t}\right)$$

$$K_2 = \frac{S_i - \left(S_t + \frac{\Delta t}{2} S_1\right)}{\tau} - \frac{\mu_m}{Y}\left(\frac{\left(S_t + \frac{\Delta t}{2} S_1\right)\left(X_t + \frac{\Delta t}{2} X_1\right)}{K_s + \left(S_t + \frac{\Delta t}{2} S_1\right)}\right)$$

$$K_3 = \frac{S_i - \left(S_t + \frac{\Delta t}{2} S_2\right)}{\tau} - \frac{\mu_m}{Y}\left(\frac{\left(S_t + \frac{\Delta t}{2} S_2\right)\left(X_t + \frac{\Delta t}{2} X_2\right)}{K_s + \left(S_t + \frac{\Delta t}{2} S_2\right)}\right)$$

$$K_3 = \frac{S_i - (S_t + \Delta t\, S_3)}{\tau} - \frac{\mu_m}{Y}\left(\frac{(S_t + \Delta t\, S_3)(X_t + \Delta t\, X_3)}{K_s + (S_t + \Delta t\, S_3)}\right)$$

$$S_{t+1} = S_t + \frac{\Delta t}{6}[K_1 + 2K_2 + 2K_3 + K_4]$$

For biomass:

$$\frac{dX}{dt} = \left(\frac{\mu_m \cdot S}{K_s + S} - k_d - \frac{1}{\tau}\right) X \quad \ldots\ldots\ldots\ldots\ldots\ldots\ldots\ldots (3.35)$$

Applying RK4 formulation:

$$K_1' = \left(\frac{\mu_m \cdot S_t}{K_s + S_t} - k_d - \frac{1}{\tau}\right) X_t$$

$$K_2' = \left(\frac{\mu_m \left(S_t + \frac{\Delta t}{2} S_1\right)}{K_s + \left(S_t + \frac{\Delta t}{2} S_1\right)} - k_d - \frac{1}{\tau}\right) \left(X_t + \frac{\Delta t}{2} X_1\right)$$

$$K_3' = \left(\frac{\mu_m \left(S_t + \frac{\Delta t}{2} S_2\right)}{K_s + \left(S_t + \frac{\Delta t}{2} S_2\right)} - k_d - \frac{1}{\tau}\right) \left(X_t + \frac{\Delta t}{2} X_2\right)$$

$$K_4' = \left(\frac{\mu_m (S_t + \Delta t\, S_3)}{K_s + (S_t + \Delta t\, S_3)} - k_d - \frac{1}{\tau}\right) (X_t + \Delta t\, X_3)$$

$$X_{t+1} = X_t + \frac{\Delta t}{6} [K_1' + 2 K_2' + 2 K_3' + K_4']$$

3.5 Kinetic parameter estimation

Kinetic parameter estimation is a critical component of anaerobic digestion (AD) modeling. It involves determining the numerical values of parameters that define the kinetic models governing microbial growth, substrate degradation, and product formation. Accurate estimation of kinetic parameters allows engineers, researchers, and process designers to simulate digestion performance, design bioreactors, predict methane yields, and implement control strategies (Lee et al., 2017a). In anaerobic systems—where complex consortia of microorganisms work synergistically across multiple metabolic stages—the importance of parameter estimation becomes evenmore pronounced. Given the non-linear and often inhibited behavior of AD systems, parameter estimation must be carefully performed using both experimental and computational tools. This chapter elaborates on the methodology, theoretical framework, tools, and challenges associated with kinetic parameter estimation in AD, focusing on its indispensable role in process understanding, optimization, and modeling.

Kinetic parameters are intrinsic to all mathematical models of anaerobic digestion. Whether using simple first-order decay models or comprehensive frameworks like the Anaerobic Digestion Model No. 1 (ADM1), precise knowledge of kinetic parameters is necessary for: Predictive modeling of substrate degradation and biogas production. Design and scaling of digesters (e.g., sizing, retention time calculations). Optimization of operating parameters (e.g., loading rate, pH, temperature). Sensitivity analysis and risk assessment. Model calibration for digital twins and real-time control. Without accurate parameter values, even the most sophisticated model can yield misleading or unreliable results.

Kinetic parameter estimation is the backbone of anaerobic digestion modeling. It bridges the gap between theoretical models and practical application, allowing for accurate simulation, design, and optimization of digesters. While challenges remain in capturing the complexity of microbial interactions and environmental variability, ongoing improvements in experimental methods, statistical tools, and computational power continue to enhance the reliability and utility of kinetic modeling in anaerobic systems. As AD systems diversify in scale, substrate, and technological complexity, robust parameter estimation methodologies will remain central to advancing both the science and engineering of sustainable waste treatment and bioenergy production.

3.6 Application of biokinetic modeling in performance prediction

Biokinetic modeling entails the application of mathematical equations to describe the rate of microbial growth, substrate consumption, product formation, and decay based on reaction kinetics. These models allow engineers and scientists to simulate reactor behavior under varying operational scenarios and thereby predict performance outcomes such as biogas yield, volatile solids reduction, and stability indicators like volatile fatty acids (VFAs) and ammonia concentrations. At the heart of biokinetic modeling lies a set of differential equations derived from empirical or mechanistic kinetic expressions. Common models includethe Monod model for substrate-limited microbial growth, the Contois model for biomass-dependent kinetics, and first-order models for substrate degradation. These expressions are integrated into mass balance frameworks to model the dynamic behavior of theAD system over time.

Biokinetic models are used to forecast how an AD system will respond to changes in operational conditions, feedstock characteristics, and environmental parameters. Their applications in performance prediction are diverse and essential to the engineering design and management of digesters. One of the most direct applications of biokinetic modeling is the prediction of substrate degradation efficiency and biogas yield. By modeling the hydrolysis and fermentation steps using first-order or Monod kinetics, one can estimate the fraction of volatile solids (VS) converted into methane and carbon dioxide. For example, the hydrolysis rate constant (k_h) significantly influences how quickly particulate matter like lignocellulosic biomass or food waste is broken down. Coupled with methanogenic kinetics (μ_{max}, K_s), the model predicts cumulative biogas production over time, peak production rates, and the lag phase duration. Models such as the modified

Gompertzmodel or first-order kinetics are often used in batch setups, while continuous systems use CSTR-based mass balances combined with kinetic expressions.

Chapter 4. Artificial Intelligence (AI)-based methods in Anaerobic Digestion

This chapter explores how AI techniques are transforming the monitoring, control, and optimization of anaerobic digestion processes. From forecasting biogas production to detecting microbial shifts, AI offers powerful tools to overcome traditional modeling limitations.

4.1 General Introduction to AI

AI refers to the development of computer systems capable of performing tasks that traditionally require human intelligence, such as learning, reasoning, problem-solving, and decision-making. Unlike conventional algorithmic approaches, which rely on explicitly programmed rules, AI systems learn patterns and relationships directly from data. This data-driven paradigm allows AI to adapt to complex, nonlinear, and dynamic systems where purely mechanistic models may be inadequate or difficult to calibrate.

The past decade has seen rapid advances in AI across a range of fields, driven by improvements in computational hardware (e.g., GPUs, TPUs), the availability of large datasets, and breakthroughs in machine learning algorithms, particularly deep learning. In environmental engineering, AI has moved from an emerging novelty to a core toolset for process optimization, predictive maintenance, and real-time monitoring. Anaerobic digestion research has followed this trend, increasingly integrating AI-based approaches to enhance system understanding and operational efficiency.

The advantages of AI for AD applications stem from several characteristics of the technology:

- Nonlinear modeling capability: AI models can capture intricate relationshipsbetween input variables (e.g., feedstock composition, temperature, microbial dynamics) and outputs (e.g., methane yield) without assuming linearity.

- Data adaptability: AI can handle heterogeneous data types, including time-series sensor readings, genomic data, and imaging datasets, integrating them into unified predictive frameworks.
- Real-time learning potential – Certain AI methods can adapt to new data as it becomes available, enabling proactive operational control.
- Robustness to noise and missing values –With appropriate preprocessing and training, AI can extract useful patterns from imperfect datasets common in industrial AD systems.

In AD, the deployment of AI complements, rather than replaces, existing process models. While mechanistic and kinetic models remain essential for mechanistic understanding and hypothesis testing, AI can fill gaps where model parameters are unknown, system dynamics are highly variable, or rapid predictions are required without extensive computation.

4.1.1 Basics of AI

At its core, AI encompasses a spectrum of computational methods, from traditional machine learning (ML) algorithms such as linear regression, decision trees, and support vector machines to modern deep learning architectures like convolutional neural networks (CNNs) and transformers.

Key components of AI systems include:

- **Data**: The foundation for any AI model, encompassing historical and real-time measurements.
- **Features**: Processed representations of raw data that highlight relevant information for the model.
- **Model architecture**: The mathematical structure defining how input features are transformed into predictions.
- **Training process**: Iterative adjustment of model parameters to minimize prediction error on historical data.
- **Evaluation and validation**: Assessment of model generalizability using unseen data.

In the context of AD, AI models typically ingest one or more types of datasets

- **Physicochemical data**: Time-series measurements of pH, volatile fatty acids (VFAs), alkalinity, gas flow, and methane content.
- **Microbial community data**: Taxonomic profiles from sequencing (qualitative) or qPCR counts (quantitative).
- **Image-based data:** Microscopy and spectroscopy images of sludge granules, biofilms, or feedstock.
- **Operational metadata**: Information on temperature regimes, reactor type, and feedstock origin.

The flexibility of AI methods allows them to combine these disparate datasets into multi-modal predictive frameworks, enabling richer insights than single-source data analysis. Detailed descriptions of each data type are provided later in Chapter 5.

4.1.2 AI Approach in Anaerobic Digestion

Early applications of AI in AD systems date back to the late 1990s and early 2000s, focusing mainly on empirical modeling of methane yield using small sets of process variables and statistical learning methods. Over time, the scope and sophistication of AI applications in AD have expanded significantly, driven by improvements in sensor technology, sequencing capabilities, and computational resources.

Some landmark applications include:

- **Biogas production forecasting:** Models predicting daily or weekly methane yield from historical sensor readings and feedstock composition. Both shallow ML models (e.g., Random Forest, SVR) and deep architectures (e.g., LSTM, Transformer) have been employed to capture temporal dependencies and seasonal variations.
- **Process stability monitoring:** AI classifiers have been trained to distinguish between stable and unstable

operational states using physicochemical indicators (pH, VFAs, alkalinity ratios) and microbial shifts.

- **Microbial community interpretation:** Machine learning has been used to link microbial composition to process performance, identifying taxa that serve as bio indicatorsof inhibition or optimal operation.
- **Feedstock characterization:** Hyperspectral imaging combined with AI-based regression models enables rapid, non-destructive estimation of feedstock composition, supporting real-time load adjustments.
- **Fault detection and diagnosis**: Anomaly detection algorithms can identify deviations from normal operation before catastrophic process failures occur, allowing preemptive interventions.
- **Process optimization**: AI-driven optimization frameworks suggest operational adjustments (e.g., OLR, mixing rate) to maximize yield while maintaining stability.

Recent research has shifted toward hybrid modeling combining mechanistic kinetic models (e.g., ADM1) with AI components. This allows AI to capture unknown nonlinearities while preserving the interpretability of mechanistic models. Furthermore, multi-modalAI systems integrating sensor, genomic, and image data are emerging, promising unprecedented insight into AD system behavior.

4.1.3 Common AI Tasks in Anaerobic Digestion Systems

AI applications in AD can be grouped into broad task categories, each serving specific operational and research goals. These tasks are generally implemented under three main learning paradigms:

- **Supervised learning:** Models learn from labeled datasets, where each input is paired with a known output. This is common in **prediction** and **classification** tasks in AD, such as forecasting methane yield or identifying stable vs. unstable operation.

- **Unsupervised learning:** Models uncover patterns or groupings in unlabeled data. This is especially valuable when explicit target labels are unavailable, for example in **clustering** microbial community profiles or discovering distinct operating regimes.
- **Semi-supervised learning:** Models utilize a small set of labeled data together with a larger pool of unlabeled data. This is particularly relevant in AD, where comprehensive labeled datasets (e.g., process instability events) are rare, but abundant unlabeled operational data is collected continuously. In AD applications, semi-supervised methods are often employed for tasks such as process **anomaly detection**, where labeled examples of abnormal states are rare but normal operating data is plentiful.

The choice of paradigm influences the model design, data requirements, and evaluation metrics, with further details on learning objectives provided in Section 4.3 Model Training and Validation.

4.1.3.1 Prediction

Prediction tasks involve forecasting future values of key process indicators, such as:

- **Methane yield and biogas volume** – Predicting daily, weekly, or seasonal production for energy planning.
- **Gas composition** – Estimating methane and CO_2 fractions to ensure gas quality.
- **System stability metrics** – Forecasting VFA/alkalinity ratios or pH drops to anticipate instability.

Models for prediction often leverage time-series learning architectures (e.g.,LSTM, GRU, Transformer) capable of handling sequential dependencies. Detailed description of each model will be introduced in Section 4.2.

4.1.3.2 Classification

Classification tasks assign operational states or samples to predefined categories, for example:

- **Stable vs. unstable operation** – Based on physicochemical thresholds or AI-learned patterns.
- **Inhibited vs. non-inhibited states** – Detecting ammonia toxicity, organic overloading, or sulfide inhibition.
- **Feedstock type identification** – Classifying incoming substrates for automatic process adjustments.

Supervised learning algorithms such as support vector machines, gradient boosting, and deep neural networks are frequently used here.

4.1.3.3 Clustering

Clustering is an unsupervised learning task used to group similar data points without predefined labels:

- **Grouping operational regimes** – Identifying natural clusters in historical process data that correspond to distinct operating strategies or seasonal trends.
- **Microbial profile clustering** – Grouping microbial community samples to detect shifts or recoveries after disturbances.
- **Process phase separation** – Differentiating between startup, steady-state, and overload phases.

Methods such as k-means, hierarchical clustering, and self-organizing maps (SOMs) are popular choices.

4.1.3.4 Anomaly Detection

Anomaly detection aims to identify observations that deviate significantly from normal operating patterns:
- **Sensor fault detection** – Identifying outliers caused by sensor drift or malfunction.
- **Process upset detection** – Recognizing abnormal changes in VFAs, gas output, or microbial abundance.

- **Biofilm/granule deterioration detection** – Using image-based models to spot abnormal morphology linked to reduced performance.

Techniques range from statistical thresholds to more advanced methods like autoencoders and isolation forests.

4.2 Description of Model Architectures

In this section, we describe the most widely used model architectures in AI-based AD research. Section 4.2.1 focuses on machine learning algorithms, which include well-established, often interpretable methods such as Linear Regression, Support Vector Machines, Naïve Bayes classifiers, Decision Trees, and Ensemble Learning Models. These models are typically well-suited for structured tabular data, offering clear interpretability and modest computational demands. Section 4.2.2 addresses neural network architectures, including Multi-Layer Perceptrons (MLP), Convolutional Neural Networks (CNN), Recurrent Neural Networks (RNN), and Transformer models. Neural networks excel at modeling complex, nonlinear relationships and are especially powerful when dealing with high-dimensional, multi-modal AD datasets (e.g., combining time-series sensor readings, microbial data, and image data).

By exploring both conventional and deep learning model architectures, this chapter provides the foundation for understanding how AI tools can be selected and tailored to specific AD research objectives, balancing predictive performance, interpretability,and computational efficiency.

4.2.1 Machine Learning algorithms

Machine learning algorithms form the foundation of data-driven modeling, where the objective is to learn predictive patterns from data. In supervised learning, which is the focus of this section, models are trained using input features (X) paired with corresponding labels (Y). The training process involves optimizing model parameters to minimize a loss or cost function, which quantifies the discrepancy between predicted and true values. For each algorithm discussed, we introduce the underlying principle, the specific cost function used for training, the key assumptions that guide the model's behavior, and the main advantages and disadvantages. Finally, each algorithm is illustrated with an application example in anaerobic digestion (AD) research, demonstrating how these methods can be practically implemented for tasks such as process monitoring, classification of operational states, and biogas production prediction.

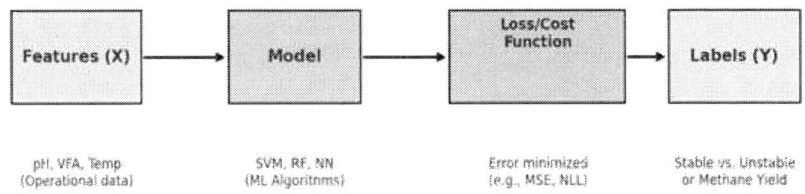

Fig 4.1 Workflow of supervised machine learning in AD research.

Features (e.g., pH, VFA, temperature) are mapped through a model to minimize a loss function and predict labels such as process stability or methane yield.

4.2.1.1 Linear regression

Linear regression is one of the most fundamental and interpretable machine learning algorithms. It models the linear relationship between a dependent variable y and one or more independent variables $x_1, x_2, \ldots, x_n$. The simplest form, simple linear regression, considers only one independent variable while multiple linear regression handles multiple predictors. In the context of AD research, a typical application would be predicting biogas production based on input substrate amount and various process parameters across different digesters.

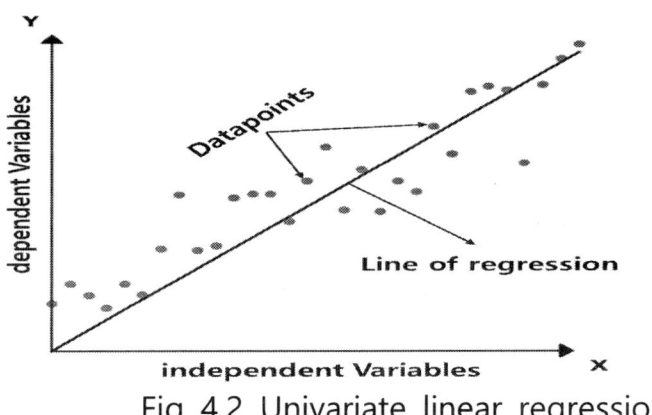

Fig 4.2 Univariate linear regression

The mathematical form of a linear regression model is:

$$y = \beta_0 + \beta_1 x_1 + \beta_2 x_2 + \cdots + \beta_n x_n + \epsilon \quad \ldots\ldots\ldots\ldots\ldots\ldots\ldots\ldots (4.1)$$

Where:
- y: the target variable (e.g. daily biogas production)
- x_i: the independent variables (e.g. influent volume, input VS, effluent pH, residual VFA concentration)
- β_i: the model coefficients estimated from data
- ϵ: the error term

In supervised learning, the model is trained using historical data where y (the true value) is known. The fitting process involves estimating the coefficients β_i such that the discrepancy between observed and predicted values is minimized.

The most common approach is **ordinary least squares (OLS)**, which minimizes the sum of squared residuals. The **cost function** for OLS is defined as:

$$J(\beta) = \frac{1}{m}\sum_{j=1}^{m}\left(y^{(j)} - \hat{y}^{(j)}\right)^2 \quad \cdots \cdots \cdots \cdots \cdots \cdots \cdots (4.2)$$

Where:
- m: number of training samples
- $y^{(j)}$: actual observed value for the j-th sample
- $\hat{y}^{(j)}$: predicted value for the j-th sample based on the model

By minimizing $J(\beta)$, the algorithm finds the optimal set of coefficients that best describe the relationship between inputs and outputs in the dataset.

The advantage of this model lies in its simplicity and interpretability. Each coefficient represents the weight of an independent variable, allowing researchers to directly assess which variables have the greatest influence on the target value. In addition, linear regression is computationally efficient on small to medium-sized datasets, as it requires relatively low memory and processing power, and its closed-form solution via OLS can be obtained quickly without iterative optimization. This makes it particularly suitable when data collection is expensive or when rapid prototyping is needed. However, the model assumes a strictly linear relationship between inputs and outputs, an assumption that may not hold in complex and nonlinear biological systems such as AD. Furthermore, linear regression is highly sensitive to outliers that even a few extreme data points can disproportionately affect the estimated coefficients, potentially leading to biased predictions and misleading interpretations. The key assumptions underlying linear regression include the independence of observations, a linear relationship between predictors and the response. In addition, the difference between actual value (y) and predicted value ($\hat{y}$) should have constant variance across all values of the input. This is called homoscedasticity and if the variance has increasing or decreasing trend, it's called heteroscedasticity. Violations of these assumptions can degrade model accuracy and validity.

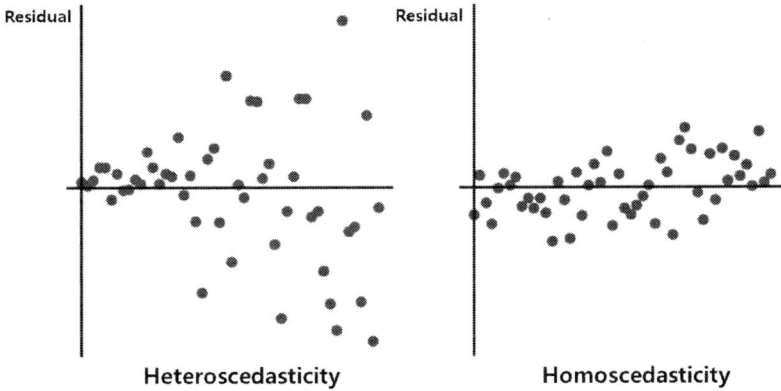

Fig 4.3 Heteroscedasticity (Left) vs Homoscedasticity (Right)

4.2.1.2 Support Vector Machines (SVM)

Support Vector Machines (SVM) are supervised learning models used for both classification and regression tasks. The fundamental idea in classification is to find an optimal hyperplane that separates data points from different classes with the maximum possible margin. In regression (Support Vector Regression, SVR), the objective is to fit a function within anacceptable deviation from the observed values. The decision boundary for classification is mathematically defined as:

$$w \cdot x + b = 0 \quad\quad\quad\quad\quad\quad\quad\quad\quad\quad (4.3)$$

where:
- ⊙ w: the weight vector perpendicular to the hyperplane,
- ⊙ x: is the input vector,
- ⊙ b: is the bias term

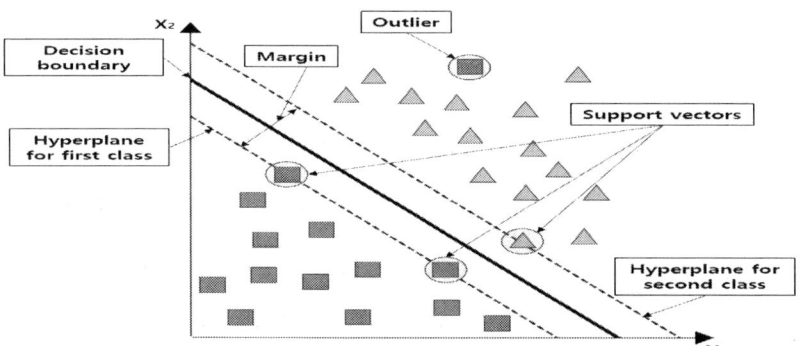

Fig 4.4 SVM illustration for binary classification problem

The figure above illustrates an example of a binary classification problem in a two-dimensional feature space (x_1, x_2). Red circles represent data points from first class and blue squares represent data points from the second class. The decision boundary (black dashed line) is the hyperplane that separates the two classes. Support vectors (circled points) are the critical data points lying closest to the decision boundary. They determine the orientation and position of the hyperplane. Two parallel lines, known as margins (solid black lines), run on either side of the decision boundary and pass through the support vectors. The margin is defined as the gap between the hyperplane for the first class and the hyperplane for the second class. SVM optimizes this margin so that it is as wide as possible, which improves generalization and classification robustness.

At first glance, one might wonder: if SVM is fundamentally about finding linear boundaries, why do we sometimes see nonlinear decision surfaces? The answer lies in the use of **kernel functions.** When data are not linearly separable, SVM employs kernel functions to project the data into a higher-dimensional space where linear separation becomes feasible. Once classification is achieved in this transformed space, the separating hyperplane can be mapped back to the original feature space. This mapping often appears as a nonlinear boundary, even though the separation in the higher-dimensional space remains linear.

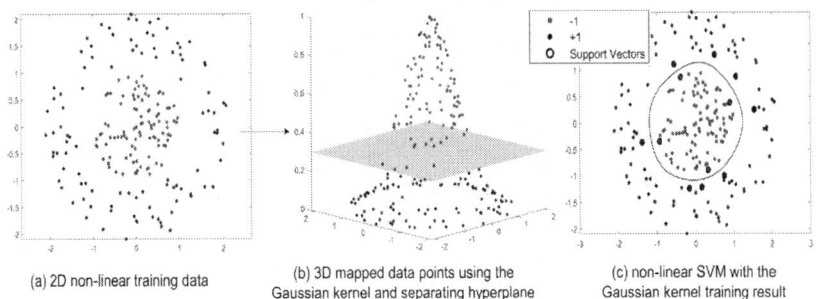

(a) 2D non-linear training data　(b) 3D mapped data points using the Gaussian kernel and separating hyperplane　(c) non-linear SVM with the Gaussian kernel training result

Fig 4.5 Nonlinear SVM classification using a Gaussian kernel. Data that are not linearly separable in two dimensions (a) become separable in a higher-dimensional feature space (b). Mapping back to the original space yields a nonlinear decision boundary (c), defined by the support vectors (Wang et al., 2024)

Commonly used kernels include:

- **Linear kernel** – no transformation, best for linearly separable data.
- **Polynomial kernel** –maps data into a higher-dimensional polynomial space.
- **Radial Basis Function (RBF)** kernel – creates complex, nonlinear decision boundaries.
- **Sigmoid kernel** – resembles a neural network activation function.

Key strengths of SVM include its effectiveness in high-dimensional feature spaces and its robustness to overfitting, especially when the number of features exceeds the number of samples. However, SVM can become computationally expensive for very large datasets, and selecting the right kerneland hyperparameters is critical for performance. The key assumption of this model is that the data can be separated by a hyperplane and observations are independent and from the same distribution.

SVM is particularly useful in AD systems where datasets are noisy, nonlinear, and require clear classification of operational states. A common application is distinguishing between stable and unstable reactor conditions. For instance, by using physicochemical variables such as pH, alkalinity, and volatile fatty acids, the data can be projected into a feature space where SVM establishes a separating hyperplane. This enables the model to classify reactor operation into "stable" or "unstable"categories with improved robustness compared to simpler linear methods.

4.2.1.3 Naïe Bayes

Naïe Bayes is a family of probabilistic classification algorithms based on Bayes'Theorem, with a strong assumption that all features are conditionally independent given the class label.

Despite this "naïe" assumption, the algorithm performs surprisingly well in many practical situations, especially with high-dimensional data.

Bayes' Theorem is expressed as:

$$P(C|X) = \frac{P(X|C) \cdot P(C)}{P(X)} \quad \dots \dots \dots \dots \dots \dots \dots \dots \dots \dots (4.4)$$

Where:
- $P(C|X)$: Posterior probability of class C given features X
- $P(X|C)$: Likelihood of features under class C
- $P(C)$: Prior probability of class C
- $P(X)$: Marginal probability of the features

For multiple features, the posterior probability can be written as:

$$P(C|X) = P(C|X_1, X_2, X_3, \cdots, X_n) = \frac{P(X_1 \cap X_2 \cap \cdots \cap X_n | C) P(C)}{P(X_1, X_2, X_3, \cdots, X_n)} \quad \dots \dots \dots (4.5)$$

In Naiive Bayes, the likelihood term $P(X|C)$ is simplified by assuming all features X_i are independent. Thus, the joint likelihood becomes:

$$P(X_1 \cap X_2 \cap \cdots \cap X_n | C) = \prod_{i=1}^{n} P(X_i | C) \quad \cdots (4.6)$$

This assumption greatly reduces model complexity and computational cost Since $P(X_1, X_2, X_3, \ldots, X_n)$ is constant for a given input, classification can be based on the following rule:

$$P(C | X_1, X_2, X_3, \cdots, X_n) \propto P(C) \prod_{i=1}^{n} P(X_i | C) \quad \cdots (4.7)$$

$$\hat{C} = \arg\max_{C} P(C) \prod_{i=1}^{n} P(X_i | C) \quad \cdots (4.8)$$

This corresponds to **Maximum A Posteriori (MAP) estimation**, where both the prior $P(C)$ and the likelihood terms $P(X_i | C)$ are estimated from the training data. In practice, $P(C)$ is often computed as the relative frequency of class $P(C)$ in the training set.

To train a Naïve Bayes model, parameters are typically estimated by maximizing the likelihood of the training data. Equivalently, this is achieved by minimizing the negative log-likelihood (NLL) cost function:

$$\mathcal{L}(\theta) = -\sum_{j=1}^{m} \log P\left(C^{(j)} | X^{(j)}; \theta\right) \quad\quad\quad (4.9)$$

Where:
- ⏵ m: the number of training samples
- ⏵ $C^{(j)}$: the true class label of sample j
- ⏵ $X^{(j)}$: the feature vector
- ⏵ θ: the model parameters

This cost function ensures that the model assigns high probability to the correct class labels during training. Since Naïve Bayes methods are a set of supervised learning algorithms for classification task, it can also be applied to classify operational states. The central assumption of Naïe Bayes is conditional independence of features given the class label. In AD applications, this implies that parameters such as VFA, alkalinity, and pH are considered independent once the operational state (stable vs. unstable) is known. While in reality these variables may be correlated, the assumption often holds wellenough to produce robust classifications, particularly when rapid decision-making is prioritized over modeling complex.

Naïve Bayes offers several advantages: it is simple to implement, computationally efficient, and works well with small datasets and high-dimensional feature spaces. It also provides probabilistic outputs, which are useful for risk assessment in AD operations. However, its main limitation is the strong independence assumption, which is often violated in real-world data where parameters are correlated (e.g., pH and alkalinity). Additionally, Naïe Bayes may underperform when the true decision boundary is highly complex, making it less suitable for nuanced predictions compared to more flexible models such as SVMs or neural networks.

4.2.1.4 Decision Trees

Decision trees are non-parametric supervised learning models used for both classification and regression tasks. They work by recursively partitioning the feature space into regions based on feature values, forming a tree-like structure of decision rules. Each internal node corresponds to a test on a feature, each branch represents the outcome of the test, and each leaf node represents a predicted class (in classification) or a numerical value (in regression). The goal of the tree-building process is to create splits that maximize the separation of classes (or minimize prediction error) at each stage.

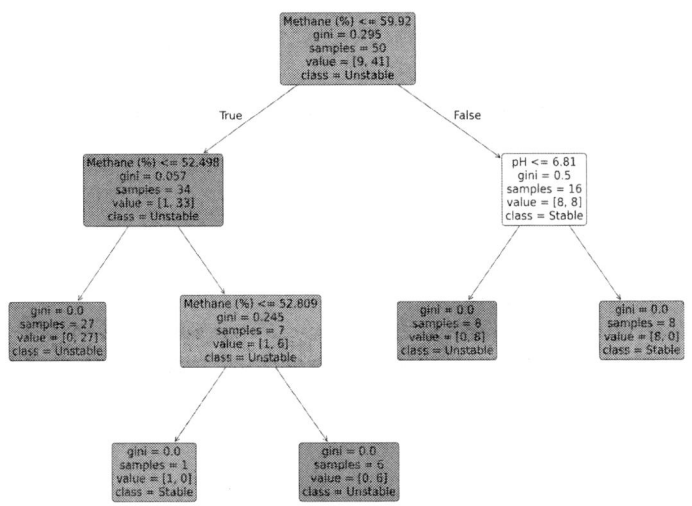

Fig 4.6 Example decision tree model for classifying anaerobic digestion stability using pH, VFA concentration, and methane yield. Internal nodes represent decision rules, while leaf nodes indicate predicted classes (Stable vs. Unstable).

The tree grows by choosing splits that minimize impurity (or maximize information gain).

Mathematically, the impurity of a split is often measured using criteria such as:

- Gini index (classification):

$$\text{Gini} = 1 - \sum_{i=1}^{K} p_i^2$$

  Where p_i is the proportion of samples belonging to class i

- Entropy (classification from information gain):

$$\text{Entropy} = -\sum_{i=1}^{K} p_i \log_2(p_i)$$

- Mean squared error (MSE) (regression):

$$\text{MSE} = \frac{1}{n} \sum_{j=1}^{n} (y_j - \hat{y}_j)^2$$

Training a decision tree involves minimizing the overall impurity or error across all nodes. The general cost function for classification can be expressed as:

$$\mathcal{L}(T) = \sum_{t \in leaves(T)} \frac{N_t}{N} \cdot H_t \quad \ldots\ldots\ldots\ldots\ldots\ldots\ldots\ldots\ldots (4.10)$$

Where:

- T: Tree
- N_t: the number of samples in leaf t
- N: the total number of samples
- H_t: the impurity of leaf t

For regression trees, H_t is typically the variance or MSE of the target values in that leaf.

Decision trees make no assumptions about the distribution of the data or linearity of relationships between features and outcomes. Instead, they assume that meaningful decision rules can be discovered by partitioning the feature space. In AD datasets, this means trees can handle heterogeneous data (e.g., categorical feedstock types and continuous variables like pH or VFAs) without requiring strong statistical assumptions.

One of the main advantages of decision trees is that they are easy to interpret and visualize, making them accessible to both researchers and operators. They can handle both numerical and categorical variables, capture nonlinear relationships, and model feature interactions without requiring complex transformations. Furthermore, they require minimal preprocessing, as there is no need for scaling or normalization of input features.

Despite these strengths, decision trees also come with limitations. They are highly prone to overfitting, especially when grown to large depths without pruning. Small variations in the dataset can lead to significantly different tree structures, making them unstable in practice. In addition, they may show bias toward features with many possible split values and, in terms of predictive performance, are often outperformed by ensemble methods such as

Random Forests and Gradient Boosting which will be introduced more in nextpart "Ensemble Learning Models".

Decision trees have been applied in anaerobic digestion to classify process states and predict performance indicators. For instance, using features such as temperature, organic loading rate (OLR), pH, and VFA concentration,a decision tree can generate interpretable rules like "if pH < 6.5 and VFA > 2000 mg/L, then the digester is unstable."This rule-based interpretability makes decision trees especially valuable for operators who require transparent and actionable insights for process monitoring and early warning systems.

4.2.1.5 Ensemble Learning Models

The core idea of ensemble learning is to establish a predictive model that combines the strengths of a collection of individual models as its base learner. Several individual models were trained on the same dataset and produced independent predictions with different learning capabilities. Those predictions are then given into ensemble models to generate a final prediction that is more robust and less prone to errors. The most widely used ensemble approaches are bagging, boosting, and stacking.

- **Bagging**(Bootstrap Aggregating) reduces variance by training multiple base learners (often decision trees) on different bootstrap samples of the dataset and averaging their predictions. Random Forest is the most well-known example.
- **Boosting** reduces bias by training models sequentially, with each new model focusing on correcting the errors of the previous ones. Popular algorithms include AdaBoost, Gradient Boosting, and XGBoost.
- **Stacking** combines the predictions of multiple heterogeneous models (e.g., SVM, decision

trees, Naïe Bayes) using a meta-learner to improve overall performance.

By leveraging the diversity of base learners, ensemble methods generally achieve higher generalization performance than single models. The most widely adopted algorithms in AD research are Random Forest, AdaBoost, Gradient Boosting, and XGBoost. These arethe brief descriptions of each model.

Random Forest (Bagging)

Random Forest constructs an ensemble of decision trees, each trained on a bootstrap sample of the training data. At each split, only a random subset of features is considered, reducing correlation among trees. Decision tree is high variance and low bias model, where any slight change in the training set may considerably alter the model performance. To overcome this limitation, each tree in Random Forest is trained on different random subset ofdata with different random subset of features to split the nodes. The average prediction from different trees is then generated as the final prediction. Therefore, Random Forest is a collection of trees that are different and independent from each other

AdaBoost

Adaptive Boosting (AdaBoost) constructs an ensemble by sequentially training weak learners, typically shallow decision trees (stumps). After each iteration, samples that were misclassified receive higher weights, so the next learner focuses more onthese difficult cases. The final model is a weighted combination of all weak learners. AdaBoost is effective for classification tasks with relatively clean datasets, but it is sensitive to noise and outliers since misclassified points are repeatedly emphasized.

Gradient Boosting

Gradient Boosting improves performance by sequentially adding models that predict the residuals (errors) of the previous learners. At each step, the new model minimizes a differentiable loss function by moving in the negative gradient direction of the error surface, hence the name "gradient boosting." This approach is highly flexible; as different loss functions can be specified for regression or classification tasks. Gradient Boosting is powerful but requires careful tuning of hyper parameters such as learning rate, number of estimators, and maximum tree depth to avoid overfitting.

XGBoost

Extreme Gradient Boosting (XGBoost) is an advanced implementation of gradient boosting designed for speed and scalability. It incorporates regularization terms in the cost function to control model complexity and prevent overfitting:

$$\mathcal{L} = \sum_{i=1}^{m} l(y_i, \hat{y}_i) + \sum_{k} \Omega(f_k), \quad \Omega(f_k) = \gamma T + \frac{1}{2}\lambda \|w\|^2 \ldots\ldots\ldots\ldots\ldots\ldots(4.11)$$

Where

- l: loss function
- $\Omega(f_k)$: the regularization term
- T: the number of leaves
- $\|w\|$: leaf weights

XGBoost supports parallel computation, efficient handling of missing values, and sparse data optimization, making it one of the most widely used boosting algorithms in large-scale applications.

Ensemble learning assumes that combining multiple weak or diverse models leads to improved predictive performance compared to a single strong learner. Random Forest assumes variance can be reduced by averaging many decorrelated trees, while boosting methods assume that sequentially correcting errors reduces bias. In AD datasets, where noise, nonlinearity, and variability are common, both strategies provide robustness by capturing different aspects of the data.

The key advantage of ensemble methods is superior predictive performance. Random Forests provide interpretability through feature importance and are robust against overfitting, while boosting methods (AdaBoost, Gradient Boosting, XGBoost) achieve state-of-the-art accuracy in many structured data problems. However, ensembles are computationally more expensive than single models, and boosting methods are more sensitive to noise and require careful tuning. Both approaches can reduce interpretability compared to simpler algorithms, which can be a limitation in operational decision-making contexts.

Ensemble models, including Random Forest, AdaBoost, Gradient Boosting, and XGBoost, have been widely applied in anaerobic digestion research for both prediction and classification tasks. They have been used to forecast methane yield, assess process stability, and identify critical operational variables such as pH, temperature, and volatile fatty acids. Due to their robustness and high predictive power, ensemble methods are increasingly preferred over single models in AD studies.

4.2.2 Neural Network Architectures

Traditional machine learning algorithms such as linear regression, SVM, Naiive Bayes, Decision Trees and ensemble models rely on feature engineering and explicit statistical assumptions (e.g. linearity, independence). These models perform well on structured tabular data, especially when the number of features is manageable and interpretability is a priority.

In contrast, neural networks, especially in their deep learning form, automatically learn hierarchical representations of data. They are particularly powerful in capturing complex, nonlinear relationships without requiring manual feature selection. Neural networks are highly adaptable and can be applied to diverse data types including:

- Microbial abundance profiles
- Spectral and microscopy images
- Time-series sensor data

While traditional ML models can be more transparent and efficient for small datasets, neural networks become more advantageous as data volume, complexity and dimensionality increase.

4.2.2.1 Multi-Layer Perceptron (MLP)

MLPs are the foundational architecture of feedforward neural networks. An MLP consists of an input layer, one or more hidden layers, and an output layer. Each layer is composed of neurons that apply a weighted sum of their inputs followed by a nonlinear activation function. The activation function is what gives the network its expressive power. It determines how the neuron's output is transformed before being passed to the next layer. Without it, even a deep network would collapse into a linear mapping, unable to capture the nonlinear dynamics that are essential in processes like AD system.

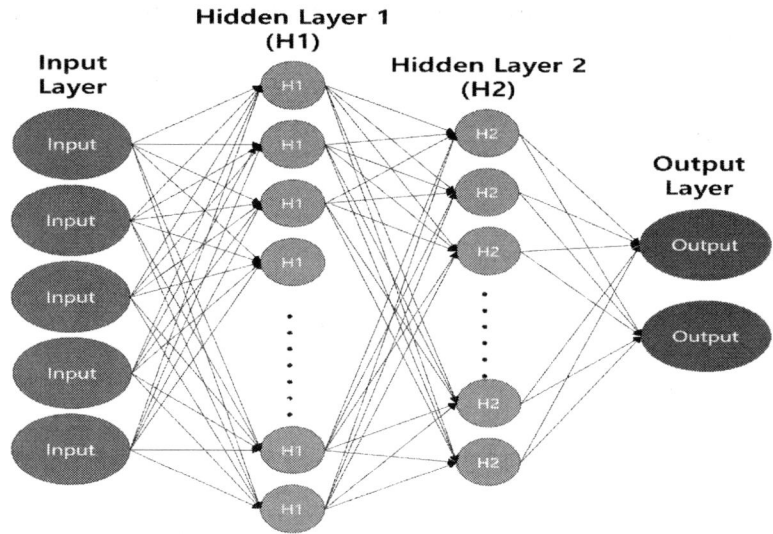

Fig 4.7 Multi-layer Perceptrons

A simple MLP with one hidden layer is mathematically described as:

$$Output = f(W_2 \cdot \sigma(W_1 \cdot X + b_1) + b_2) \quad \ldots\ldots\ldots\ldots\ldots\ldots (4.12)$$

Where:
- X: input feature (e.g. pH, VFAs, OLR)
- W_1, W_2: weight matrices for the first and second layers

- b_1, b_2: bias vectors
- σ: activation function
- f: output activation

The choice of activation function has a major impact on model performance. Common activation functions (σ) include:

- **Sigmoid:** $\sigma(x) = \frac{1}{1+e^{-x}}$ Compresses input into $(0,1)$, historically used but prone to vanishing gradients.
- **Tanh:** $tanh(x)$ Outputs in $(-1,1)$, centered around zero, often better than sigmoid for optimization.
- **ReLU (Rectified Linear Unit):** $\sigma(x) = \max(0, x)$ Widely used due to computational efficiency and mitigation of vanishing gradients, though it may suffer from "dying ReLU" when many outputs are zero.
- **Leaky ReLU / ELU:** Variants that allow small negative outputs to avoid dead neurons.

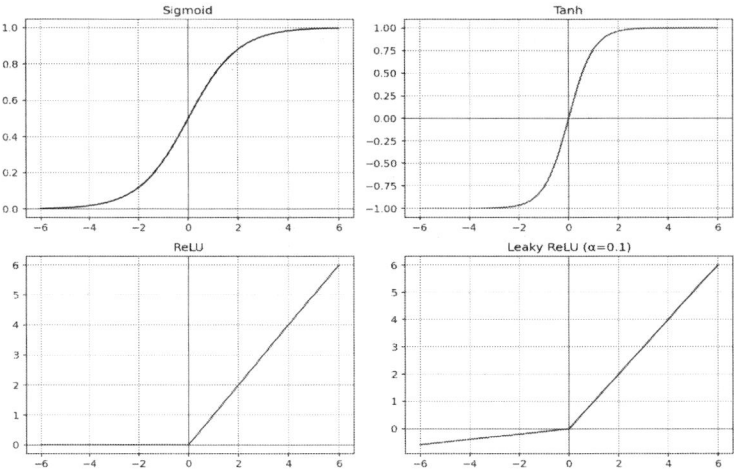

Fig 4.8 Comparison of common activation functions used in neural networks. Sigmoid and Tanh provide smooth nonlinear transformations but may suffer from vanishing gradients, while ReLU and Leaky ReLU offer sparse, efficient activations that alleviate this issue.

Output activation functions (f) tailor the network to the task:

- Identity (linear): used in regression tasks (e.g. predicting methane yield)
- Sigmoid: suited for binary classification (e.g. stable vs unstable state)

- Softmax: converts outputs into probability distributions for multi-class classification (e.g. identifying the type of input feedstock). For class i, the probability is given by:

$$P(y = i|x) = \frac{e^{z_i(x)}}{\sum_j e^{z_j(x)}} \quad \ldots \ldots \ldots \ldots \ldots \ldots \ldots \ldots \ldots \ldots (4.13)$$

Where $z_i(x)$ is the logit for class i. Softmax ensures that probabilities across all classes sum to one. In practice, when one logit dominates, softmax assigns nearly all probability to that class. Classes with weaker or constant logits, such as baseline class in this figure 4.9 (Class 3), will only receive significant probability when the stronger classes balance each other. This explains why the softmax curve for Class 3 appears relatively flat. Its probability remains low except near the point where the other two classes cancel each other out, at which point it briefly becomes competitive.

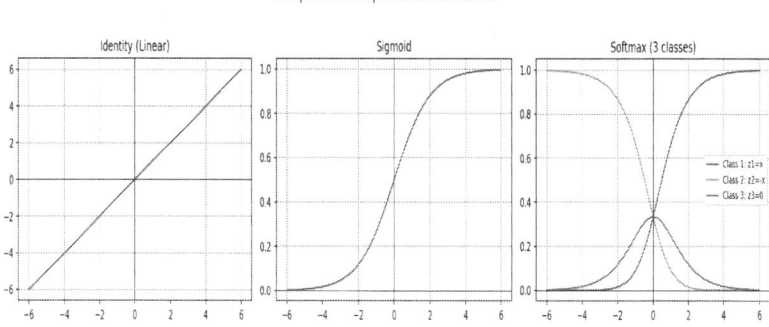

Fig 4.9 Comparison of output activation functions.

The identity function supports regression tasks, the sigmoid maps outputs to $(0,1)$ for binary classification, and the softmax distributes probabilities across multiple classes, with the baseline class (Class 3) becoming competitive only when the dominant classes balance each other.

The model learns weights by minimizing a loss function (e.g. mean squared error or cross-entropy) using backpropagation and gradient descent. Backpropagation is the process of computing gradients of the loss with respect to each weight by applying the chain rule from the output layer back through the hidden layers. These gradients indicate how much each parameter contributed to the error. Gradient descent then updates the weights in the opposite direction of the gradient, typically in small steps controlledby a learning rate, to reduce the loss. Iterating this process across many training examples gradually tunes the network parameters so that predictions better match the target outputs. The detailed mathematical calculations and step-by-step procedure of these methods will be introduced in Section 4.3.

MLPs are versatile and capable of approximating complex nonlinear functions, making them widely applicable across regression and classification tasks in AD research. They are relatively simple to implement, computationally efficient for small- to medium-sized datasets and form the conceptual foundation for more advanced architectures. However, MLPs treat all input features as independent and fully connected, which leads to many parameters and limited abilityto exploit local structures in the data. As a result, they may struggle with high-dimensional inputs such as images or spatially correlated data, where recognizing local patterns is essential. To address these limitations, more specialized architectures such as Convolutional Neural Networks (CNNs) have been developed, which introduce parameter sharing and local receptive fields to capture spatial dependencies more effectively.

4.2.2.2 Convolutional Neural Network (CNN)

Convolutional Neural Networks (CNNs)are a class of deep learning models designed to process data with a grid-like structure, such as images or other spatially organized signals. Unlike Multi-Layer Perceptron (MLP), which treat all input features equally, CNNs exploit local connectivity by applying filters (also called kernels) that slide across the input to detect localized patterns. Each filter is composed of learnable parameters (i.e. weights and biases) that are shared across spatial locations, allowing the network to efficiently capturefeatures such as edges, textures, or shapes. Multiple filters are typically used in a single convolutional layer, with each filter learning to recognize a different type of feature from the input.

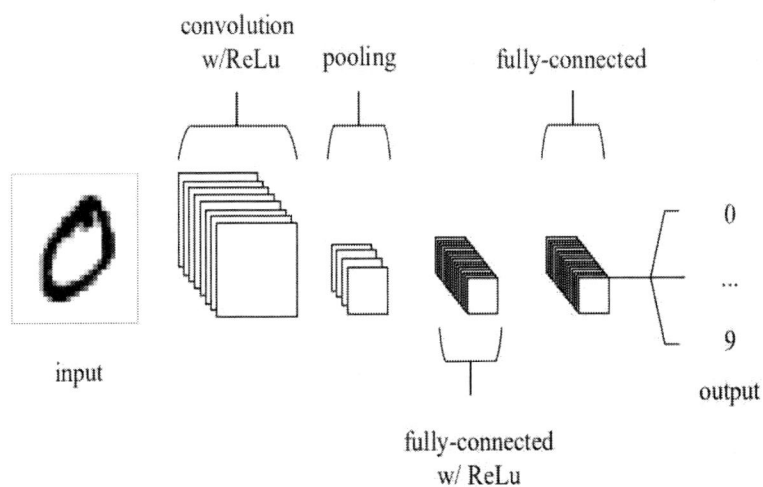

Fig 4.10 Convolution Neural Network (CNN) architecture (O'Shea & Nash, 2015)

A typical CNN consists of:

- **Convolutional layers:** Apply filters to extract feature maps (e.g., texture, edges, or spectral shifts).
- **Activation functions:** Nonlinear transformations (e.g., ReLU) applied after convolution.
- **Pooling layers:** Reduce spatial resolution (e.g., max-pooling) to generalize features and reduce computation.
- **Fully connected layers:** Interpret the extracted features for final prediction.

The input image is first represented as a numerical tensor rather than raw pixels. A grayscale image of size $m \times n$ can be expressed as a 2D matrix, where each entry corresponds to a pixel intensity value (e.g., between 0 and 255 or normalized between 0 and 1). A color image (e.g., RGB) extends this to a 3D tensor of size $m \times n \times 3$, where the last dimension represents the red, green, and blue channels.

To feed the image into a neural network such as a MLP, this tensor is flattened into a one-dimensional vector. For instance, a 28×28 grayscale image is reshaped into a vector of length 784. This flattening operation preserves all pixel values butdiscards spatial relationships among neighboring pixels. CNN address this limitation by keeping the image in its multi-dimensional form during the convolution and pooling stages, thereby maintaining local spatial structure. Flattening into a 1D vector ispostponed until later in the architecture, typically just before the fully connected (dense) layers used for classification.

　Let's go through step-by-step how the 28×28 grayscale input image is processed through CNN.

① Input representation (local receptive field)

　The image is a tensor $X \in \mathbb{R}^{28 \times 28}$. CNNs do not flatten the whole image at the start. Instead, at each spatial location (i,j) a small patch (the **receptive field**, e.g., 3×3) is extracted and **locally vectorized**: $x_{i,j} \in \mathbb{R}^9$. This "input vector" refers to the unrolled pixels of that local patch, not the entire image.

② Convolution with kernels

A convolutional kernel (filter) with weights $w \in \mathbb{R}^9$ and bias b computes

$$y_{i,j} = w^T x_{i,j} + b \quad \dots \dots \dots \dots \dots \dots \dots \dots \dots \dots (4.14)$$

Sliding this kernel across all valid (i,j) locations produce one feature map. Using F different kernels yield F feature maps.

③ Padding and Stride

When a kernel slides across the image, choices must be made about how it handles the borders and how far it moves at each step. Without adjustment, the feature map naturally shrinks as the kernel cannot fully cover pixels near the edges. To address this, padding (*P*)adds zeros around the border of the image so the kernel can also scan edge pixels and, if desired, preserve the original spatial size (e.g., keeping the feature map at 28×28). In addition, the step size of the kernel, called stride (S), controls how densely the kernel samples the image. A stride of 1 examines every neighboring pixel location, while a larger stride skips positions, reducing the spatial resolution and speeding up computation. For input $H \times W$, kernel K, padding P, stride S, the output height/width is

$$H_{out} = \left\lfloor \frac{H-K+2P}{S} \right\rfloor + 1$$

$$W_{out} = \left\lfloor \frac{W-K+2P}{S} \right\rfloor + 1 \ldots\ldots\ldots\ldots\ldots\ldots(4.15)$$

For example, 28×28 with $K=3, P=1, S=1$ gives $H_{out} = W_{out} = 28$

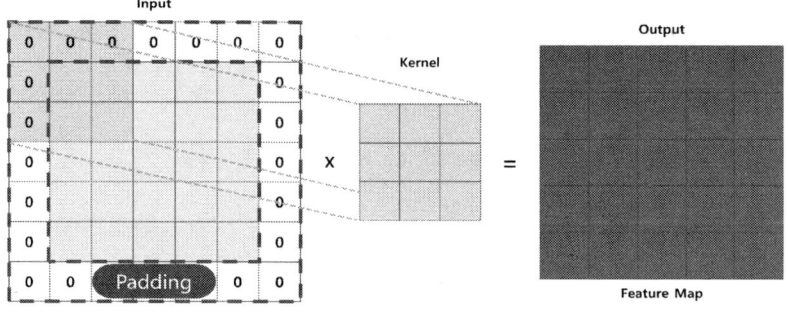

Fig 4.11 Zero padding around input image multiplied by kernel to generate feature map output

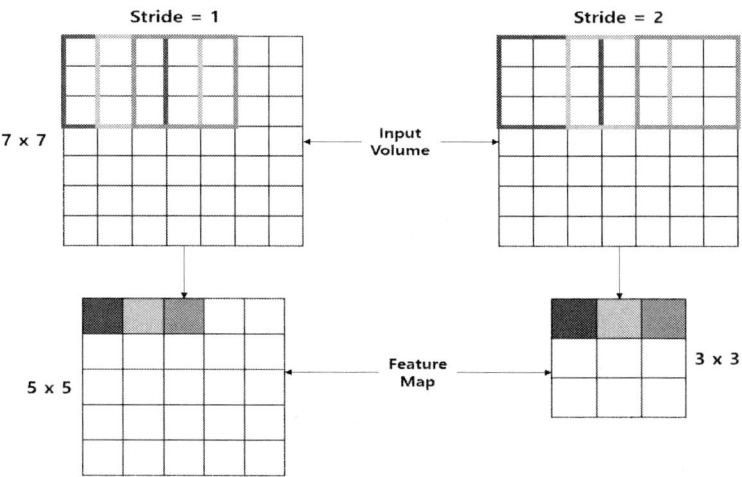

Fig 4.12 With stride 1 (left), the filter moves one

step at a time, producing a larger output feature map. With stride 2 (right), the filter skips every other position, resulting in a smaller feature map with reduced spatial resolution.

④ Feature map construction

The collection of dot products $y_{i,j}$ over all positions forms a feature map Y. With $F=32$ filters, you get 32 feature maps of size 28×28 after the first convolution.

⑤ Activation and pooling

The feature maps are then passed through a nonlinear activation function commonly ReLU to introduce nonlinearity. Then, pooling operations reduce spatial size and add translational robustness. Pooling can be done in two ways:

- ㉠ **Max pooling** selects the maximum value within a defined window (e.g., 2×2) and stride (often 2). For example, applying max pooling to a 28×28 map results in a 14×14, keeping only the strongest activation in each region.

- **Average pooling** instead computes the mean value within the pooling window, smoothing the representation and preserving overall activation trends.

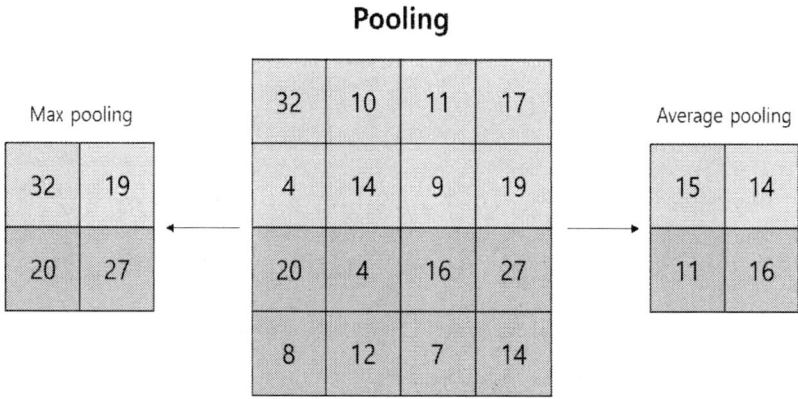

Fig 4.13 Illustration of max pooling and average pooling operation for 2×2 pooling filter. Each number represents the activation values

⑥ Deeper layers flatten fully connected prediction

Additional convolution-activation-pooling blocks are stacked to learn increasingly abstract and complex features. For example, early layers may detect edges, while deeper layers identify shapes or object parts.

After the final block, flatten the tensor of shape (F_{final}, H_{final}, W_{final}) into a 1D vector and pass it through one or more fully connected layers that integrate the extracted features. The output layer has one node per class, followed by Softmax to convert logits into a probability distribution over classes

CNNs operate under several fundamental assumptions. First, they assume local connectivity, meaning that important features can be extracted from small neighborhoods of pixels rather than requiring the entire image at once. Second, they rely on weight sharing, where the same set of filter parameters is applied across all spatial locations, reflecting the idea that a feature such as an edge or corner is equally meaningful regardless of its position in the image. Third, CNNs build on the assumption of translation invariance, which holds that the presence of a feature is more important than its exact location. Finally, CNNs assume a hierarchical feature learning process, whereby simple patterns detected in early layers (such as lines or textures) combine to form more complex structures (such as shapes or objects) in deeper layers.

These assumptions give rise to several advantages. CNNs are highly parameter-efficient, since weight sharing dramatically reduces the number of parameters compared to fully connected networks. They also preserve spatial structure, which allows them to naturally capture thetwo-dimensional relationships in images and other grid-like data. Another strength is their ability to perform automatic feature extraction, eliminating the need for manual feature engineering. In addition, CNNs demonstrate a degree of robustness to translations and small distortions, thanks to pooling operations and shared filters, and they are scalable across a wide range of input sizes, from small images such as handwritten digits to large, high-resolution datasets.

Despite these strengths, CNNs also face notable disadvantages. They are data-intensive, typically requiring large, labeled datasets to train effectively, and they are computationally demanding, often needing specialized hardware such as GPUs for efficient training. CNNs also suffer from a lackof interpretability, as the features learned by deeper layers are not always easily mapped to intuitive or domain-specific concepts. Furthermore, because of their reliance on local receptive fields, standard CNNs can struggle to capture long-range dependencies across an image. Finally, when applied to small datasets without sufficient regularization, CNNs are prone to overfitting, learning noise rather than meaningful, generalizable patterns.

4.2.2.3 Recurrent neural networks (RNNs)

Recurrent Neural Networks (RNNs) are designed to handle data that unfoldover time, making them particularly suitable for the sequential datasets generated by AD processes. Because AD is a dynamic process system, its monitoring data—such as pH, volatile fatty acids, methane concentration, or biogas yield—are naturally organized as time-series sequences. Capturing the temporal dependencies within these measurements is essential for accurate forecasting, stability assessment, and process control.

MLPs and CNNs are limited in theirability to treat sequential data. MLPs treat each input as independent, ignoring the temporal order in which values occur. CNNs can partially capture local patterns through sliding filters, but they remain constrained to fixed receptive fields and cannot flexibly model long-term dependencies across time. Consequently, both architectures fall short when the predictive task depends on trends, cycles, or lagged effects that span across multiple time steps.

RNNs overcome these limitations by introducing **recurrent connections** that allow information from previous time steps to be carried forward into the current computation. At each time step, an RNN cell receives both the current input and a hidden state vector summarizing past information. This recursive structure enables the network to maintain a form of memory, making it well-suited for modeling the temporal dynamics inherent in AD processes. By explicitly accounting for sequential order, RNNs can capture both short-term fluctuations and longer-term trends, providing a more faithful representation of time-dependent patterns compared to feedforward models.

Hidden state is calculated as:

$$h_t = \sigma(U \cdot x_t + W \cdot h_{t-1} + B) \dots\dots\dots\dots\dots\dots\dots\dots\dots (4.16)$$

Where

- h: the current hidden state
- U, W: weight matrices
- B: bias

Output is calculated as:

$$y_t = O(V \cdot h_t + C) \dots\dots\dots\dots\dots\dots\dots\dots\dots (4.17)$$

Where,

- O: activation function
- V: weight matrix
- C: bias

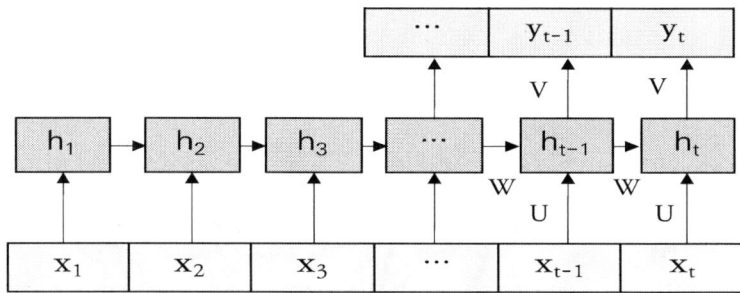

Fig 4.14 Recursive structure of RNNs (x_i: sequential input data points; h_i: hidden state at ith time-step; y_i: output at ith time-step; U, W, V: weights for x_i, h_i, y_i respectively)

RNNs share weights across time steps, enabling them to capture sequential patterns efficiently. While they are effective at modeling short-term dependencies, they face a significant challenge known as the vanishing gradient problem. During backpropagation through time, the gradients that update the network's weights can become exceedingly small as they are multiplied repeatedly across many time steps. When this occurs, the weight updates diminish toward zero, preventing the network from learning long-range dependencies. In practice, this means that standard RNNs tend to "forget"information from earlier in the sequence, making them unsuitable for tasks that require capturing patterns or influences spanning many time steps. In contrast, when gradient values are greater than one, they can grow exponentially, leading to the exploding gradient problem, where weight updates become excessively large and destabilize the training process. Together, these issues limit the effectiveness of standard RNNs for modeling long sequences in complex process data such as those produced by AD systems.

To address these issues, advanced variants such as **Long Short-Term Memory (LSTM)** networks were developed. By incorporating gating mechanisms and specialized memory cells, these architectures are designed to retain relevant information over longer time horizons while mitigating the effects of vanishing and exploding gradients.

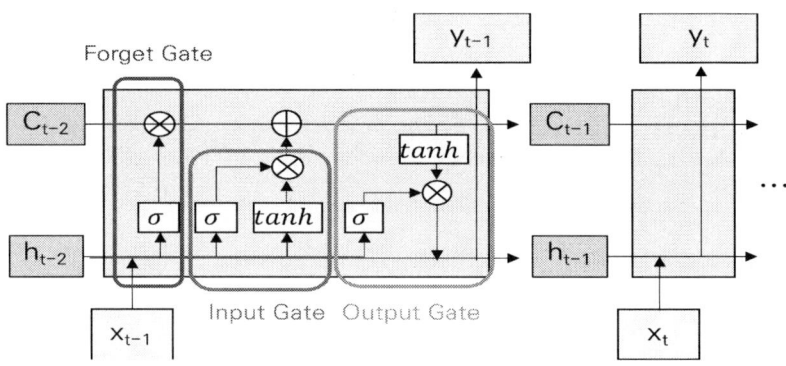

Fig 4.15 Illustration of LSTM architecture

An LSTM cell extends a standard RNN by introducing a cell state (C_t) and a system of gates that regulate information flow. These gates (forget, input, and output) control what information is remembered, updated, or discarded at each time step.

① Inputs to the LSTM cell

At time step t, the LSTM receives:

- The current input vector x_t
- The hidden state from the previous time step h_{t-1}
- The cell state from the previous time step C_{t-1}

These serve as the basis for the gating operations.

② Forget Gate (red box in Fig 4.15)

The forget gate decides which parts of the previous cell state C_{t-1} should be discarded. It applies a sigmoid activation function (σ) to a combination of h_{t-1} and x_t:

$$f_t = \sigma(Wf \cdot [h_{t-1}, x_t] + b_f) \quad \quad (4.18)$$

Here, f_t produces values between 0 and 1. A value near 0 means "forget this information," while a value near 1 means "keep it."

③ Input Gate (blue box in Fig 4.15)
The input gate controls how much new information will be added to the cell state. It has two components:

- ⓘ A **sigmoid gate** that decides which values will be updated:

$$i_t = \sigma(W_i \cdot [h_{t-1}, x_t] + b_i) \quad \cdots \cdots (4.19)$$

- ⓘ A **tanh layer** that generates candidate values $\tilde{C}_t$:

$$\tilde{C}_t = \tan h(W_C \cdot [h_{t-1}, x_t] + b_C) \quad \cdots \cdots (4.20)$$

The product $i_t \cdot \tilde{C}_t$ determines how much of this new candidate information flows into the cell state.

④ Cell state update

The cell state C_t is updated by combining the old state and the new candidate:

$$C_t = f_t \cdot C_{t-1} + i_t \cdot \tilde{C}_t \quad\quad\quad\quad (4.21)$$

This equation ensures that the LSTM selectively retains past information (via f_t) while incorporating new information (via i_t).

⑤ Output gate (green box in Fig 4.15)

The output gate determines what information is passed on as the hidden state h_t, which also serves as the output y_t.

- First, a sigmoid function decides which parts of the updated cell state contribute to the output:

$$o_t = \sigma(W_o \cdot [h_{t-1}, x_t] + b_o) \ldots\ldots\ldots\ldots\ldots\ldots\ldots\ldots\ldots (4.22)$$

ⓑ Then, the cell state is squashed through a tanh function and multiplied with o_t:

$$h_t = o_t \cdot \tanh(C_t) \ldots\ldots\ldots\ldots\ldots\ldots\ldots\ldots\ldots (4.23)$$

This gives the final hidden state, balancing long-term memory (cell state) with short-term context (hidden state).

⑥ Outputs
- ⓑ C_t: the updated cell state carrying long-term memory
- ⓑ h_t: the hidden state, also used as the output at this time step

The LSTM architecture solves the vanishing gradient problem of standard RNNs by introducing a carefully regulated memory mechanism.

- The **forget gate** removes irrelevant past information.

- The **input gate** writes useful new information into memory.

- The **output gate** determines what information is revealed at each step.

By controlling information flow through these gates, LSTMs can learn dependencies spanning dozens or even hundreds of time steps, making them especially useful for modeling sequential processes like the time-series dynamics of anaerobic digestion systems.

The design of LSTMs rests on several key assumptions. First, sequential data often contain long-term dependencies that cannot be captured by shallow or memory-less models. Second, not all past information is equally relevant; therefore, a mechanism is required to selectively remember or forget information. Third, the assumption of gated information flow suggests that a recurrent architecture can dynamically regulate memory updates and outputs at each time step. These principles underpin the gating mechanism of LSTMs, allowing them to adaptively control which signals influence the evolving memory.

LSTMs are particularly powerful in handling sequential tasks where long-term dependencies are critical. By mitigating the vanishing gradient problem, they can retain information across many time steps, making them well suited for gated information flow. The gating system provides flexibility, enabling LSTMs to filter noise and focus on relevant temporal signals. In the context of AD research, this means they can integrate fluctuations in monitoring variables such as volatile fatty acids, pH, or methane yield over extended horizons, improving the prediction of process stability and performance.

Despite their strengths, LSTMs have notable limitations. They are computationally expensive, as the recurrent structure and gating operations significantly increase training time compared to simpler models. Their parameter-heavy architecture makes them prone to overfitting, especially with small datasets. Moreover, while LSTMs address short- and long-term dependencies, they still struggle with very long sequences where dependencies span hundreds or thousands of time steps. Interpretability remains another drawback: although gates provide some intuition, it is difficult to map the internal memory dynamics directly to domain-specific phenomena in AD systems.

The limitations of LSTMs, particularly their sequential nature, reveal a deeper challenge: training is inherently time-dependent, as each step relies on the previous one, making parallelization inefficient. This sequential bottleneck slows learning on long datasets and prevents scaling to massive data streams. To overcome these issues, the Transformer architecture was introduced. By discarding recurrence altogether and relying instead on self-attention mechanisms, Transformers capture both short- and long-range dependencies in parallel. This innovation not only accelerates training but also improves the modeling of complex dependencies in sequential data, making them an especially promising approach for handling the non-stationary and high-dimensional time-series typical of AD systems.

4.2.2.4 Transformer

The Transformer architecture was introduced in 2017 with the landmark paper "Attention Is All You Need" by Vaswani et al. It was developed in response to the shortcomings of recurrent architectures such as LSTMs, which, despite their improvements over vanilla RNNs, still suffered from sequential training bottlenecks and difficulty modeling long-range dependencies. By replacing recurrence with a purely attention-based mechanism, the Transformer enabled parallel training over sequences and opened the door to scaling models to unprecedented sizes. This architectural innovation marked a turning point in deep learning for sequential data, leading to state-of-the-art performance in natural language processing, time-series forecasting, and many other domains.

At its core, the Transformer consists of two major components: an **encoder** and a **decoder**. The encoder is responsible for processing the input sequence and building contextualized representations of each element, while the decoder generates the output sequence (for example, in machine translation, the encoder reads the source language and the decoder produces the target language). Both encoder and decoder are built from stacked layers that include two critical modules: the **multi-head self-attention mechanism** and **position-wise feed-forward networks**, combined with residual connections and normalization layers. Positional encoding is also introduced to retain information about the order of elements, since the architecture itself does not rely on recurrence.

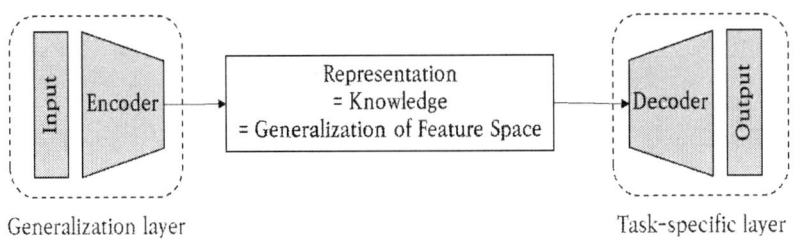

Fig. 4.16 Encoder and decoder architecture

The Transformer's self-attention mechanism plays a crucial role in the decoding process by quantifying the importance of different time steps within the encoder's contextual representations. The self-attention mechanism computes relationships between time steps using three learned matrices: Query (Q), Key (K), and Value (V). These matrices are derived from the input dataset and enable the model to assign different levels of importance to past observations. The attention score Z is calculated as follows:

$$Z = \text{Softmax}\left(\frac{QK^T}{\sqrt{d}}\right) V \ (d: dimension\ of\ key\ vectors)\ \dots\dots\dots\dots\dots\dots\dots\dots\dots\dots..(4.24)$$

The Transformer employs a **multi-head self-attention mechanism** to enhance the model's ability to capture complex temporal dependencies. In this mechanism, multiple "heads" operate in parallel, each computing independent attention scores using different learned weight matrices for Q, K, V. Each head captures distinct aspects of the input sequence, allowing the model to attend to different time steps and patterns simultaneously. By aggregating the outputs from multiple attention heads, the model can better capture short- and long-term dependencies, improving forecastingperformance. The final attention output is obtained by concatenating the outputs of all heads and passing them through a linear transformation. This multi-head mechanism allows the Transformer to learn diverse relationships within the time series data, making it particularly effective for multi-step forecasting tasks.

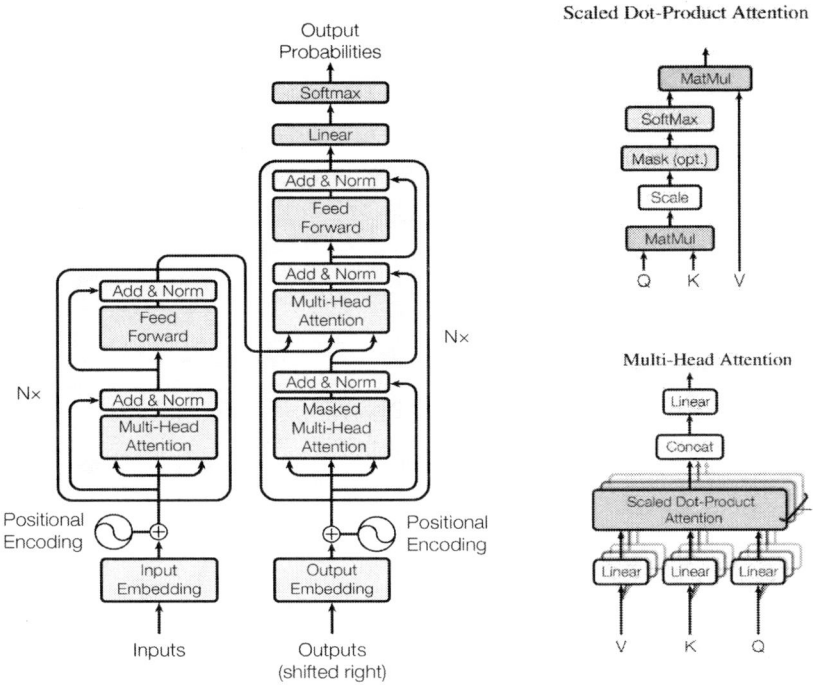

Fig. 4.17 Illustration of Transformer architecture (left); scaled dot-product attention (upper right); multi-head attention (below right) (Vaswani et al., 2023)

In this way, the Transformer architecture eliminates the sequential bottleneck inherent in recurrent models, achieves superior scalability through parallelization, and captures dependencies across any distance with remarkable efficiency. These characteristics make it not only a powerful alternative to RNN-based approaches but also a natural fit for modeling the **complex, non-stationary time-series data** encountered in anaerobic digestion systems.

Let's go through step-by-step how a Transformer works on time-series data. We will denote sequence length as T, feature dimension d_{model}, heads h, per-head key/query dimension $d_k = d_{model}/h$

① Inputs and positional encoding

Given an input sequence $x_{1:T}$ (e.g. multivariate time-series windows), each step is embedded to $e_t \in R^{d_{model}}$. Because self-attention is permutation-invariant, we add positions:

$$z_t^{(0)} = e_t + p_t, \quad t = 1, \dots, T \quad \dots\dots\dots\dots\dots\dots\dots\dots(4.25)$$

A common sinusoidal positional encoding is

$$PE(t, 2i) = \sin\left(t/10000^{\frac{2i}{d_{model}}}\right) \quad \ldots\ldots\ldots\ldots\ldots\ldots\ldots\ldots\ldots (4.26)$$

$$PE(t, 2i+1) = \cos\left(t/10000^{\frac{2i}{d_{model}}}\right) \quad \ldots\ldots\ldots\ldots\ldots\ldots\ldots\ldots (4.27)$$

② Scaled dot-product attention (single head)

For a sequence matrix $Z \in \mathbb{R}^{T \times d_{model}}$

$$Q = ZW_Q, K = ZW_K, V = ZW_V,$$

$$W_Q, W_K, W_V \in R^{d_{model} \times d_k}$$

Attention weights and outputs are

$$A = \text{softmax}\left(\frac{QK^T}{\sqrt{d_k}} + M\right), \quad \ldots\ldots\ldots\ldots\ldots\ldots\ldots\ldots (4.28)$$

$$\text{Attn}(Q, K, V) = AV \quad \ldots\ldots\ldots\ldots\ldots\ldots\ldots\ldots\ldots (4.29)$$

$M \in R^{T \times T}$ is an additive mask; for encoder self-attention it is typically all zeros; for causal or decoder self-attention, $M_{i,j} = -\infty$ when $j > i$

③ Multi-head self-attention

Compute h parallel heads, then concatenate:

$$\text{head}_i = \text{Attn}\left(ZW_Q^{(i)}, ZW_K^{(i)}, ZW_V^{(i)}\right), \dots\dots\dots\dots\dots\dots\dots\dots\dots (4.30)$$

$$\text{MHA}(Z) = \text{Concat}(\text{head}_1, \dots, \text{head}_h)\, W_O \dots\dots\dots\dots\dots\dots\dots (4.31)$$

$$W_O \in \mathbb{R}^{(hd_k) \times d_{model}}$$

④ Residual connection + layer normalization

Each sublayer is wrapped with residual add and normalization:

$$\tilde{Z} = \text{LayerNorm}(Z + \text{MHA}(Z)) \dots\dots\dots\dots\dots\dots\dots\dots (4.32)$$

LayerNorm operates feature-wise:

$$\text{LayerNorm}(x) = \gamma \odot \frac{x - \mu}{\sqrt{\sigma^2 + \varepsilon}} + \beta \dots\dots\dots\dots\dots\dots\dots (4.33)$$

⑤ Position-wise feed-forward network (FFN)

A two-layer MLP applied at each position:

$$\text{FFN}(u) = W_2\, \sigma(W_1 u + b_1) + b_2, \quad\quad\quad (4.34)$$

With σ typically ReLU or GELU, $W_1 \in \mathbb{R}^{d_{model} \times d_{ff}}, W_2 \in \mathbb{R}^{d_{ff} \times d_{model}}$

Add residual and LayerNorm again:

$$Z' = \text{LayerNorm}\left(\tilde{Z} + \text{FFN}(\tilde{Z})\right). \quad\quad\quad (4.35)$$

⑥ Encoder stack

The encoder repeats MHA Add&Norm FFN Add&Norm for L layers:

$$\boldsymbol{H}_{enc} = \text{Encoder}_L\left(\boldsymbol{Z}^{(0)}\right) \quad\quad\quad (4.36)$$

⑦ Decoder stack (seq-to-seq / autoregressive settings)

The decoder has two attentions per layer: masked self-attention over past targets, then cross-attention over encoder outputs:

$$\tilde{Y} = \text{LayerNorm}(Y + \text{MHA}(Y)), \quad\quad\quad\quad\quad (4.37)$$

$$\hat{Y} = \text{LayerNorm}(\tilde{Y} + \text{MHA}_{\text{cross}}(\tilde{Y}, H_{enc}, H_{enc})) \quad\quad (4.38)$$

$$H_{dec} = \text{LayerNorm}\left(\hat{Y} + \text{FFN}(\hat{Y})\right) \quad\quad\quad\quad (4.39)$$

In many time-series forecasters, the decoder is omitted; a projection head maps H_{enc} to future horizons.

The Transformer model is built on the assumption that sequential data can be effectively represented through attention mechanisms rather than relying on recurrent or convolutional structures. By allowing the model to compute pairwise dependencies betweenall elements of a sequence, it assumes that global contextual information is more critical for learning than strictly local or sequential connections. This assumption enables the model to capture long-range dependencies without degradation, a feature particularly valuable for time series forecasting in anaerobic digestion systems where process instabilities may be influenced by conditions that occurred far earlier in the sequence.

The foremost advantage of the Transformer architecture lies in its ability to overcome the vanishing gradient limitations of recurrent models, enabling efficient learning of long-term dependencies. Its parallelizable structure significantly reduces training time compared to sequential RNNs and LSTMs, making it scalable for large datasets. Furthermore, the self-attention mechanism provides interpretability by highlighting which time steps or features contribute most to predictions, a critical property in AD research where understanding causal relationships is as important as prediction accuracy. Multi-head attention also enhances model robustness by allowing multiple perspectives on the same input, improving the generalization of the model across diverse operational datasets. In addition, the Transformer has strong representation learning capabilities: by projecting heterogeneous and high-dimensional inputs into a shared latent space, it can capture underlying structures and correlations in complex AD datasets. This property is particularly valuable when integrating multimodal information such as physicochemical, microbial, and time-series data, thereby enabling transfer learning and adaptability across different reactor systems.

Despite its strengths, the Transformer model is not without drawbacks. Its reliance on pairwise comparisonsbetween all input tokens leads to quadratic computational and memory complexity, which can be prohibitive for very long time series common in industrial applications. The model also requires large amounts of data to train effectively, making it less suited to data-scarce environments unless combined with pretraining or transfer learning approaches. In addition, while attention weights offer some interpretability, they do not always directly translate into clear causal insights, which can limit the transparency of the model when applied to sensitive decision-making in process control. These limitations highlight the need for adaptation and careful design when applying Transformer models in anaerobic digestion research.

4.3 Training and Validation of Models

Training and validating AI models are central steps in building reliable tools. The training process enables a model to learn patterns from data, while validation ensures that the learned knowledge generalizes to new, unseen conditions. Without careful training and validation, AI models risk overfitting (i.e. capturing noise instead of meaningful trends) or underfitting, failing to capture essential process dynamics. In AD applications, where data can be noisy, heterogeneous, and time-dependent, the rigor of training and validation directly determines the credibility of model predictions such as methane yield, process stability, or microbial dynamics.

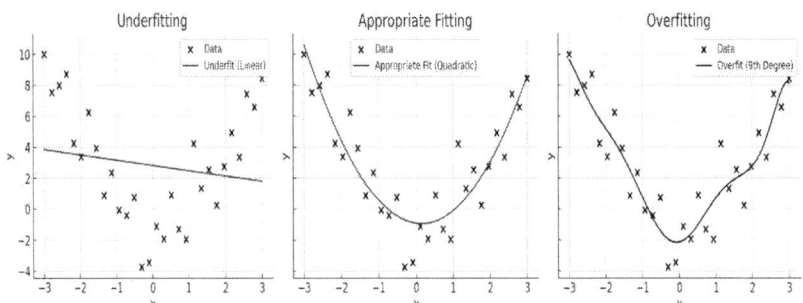

Fig. 4.18 Illustration of model fitting scenarios. Underfitting occurs when a model is too simple to capture the underlying trend of the data. Appropriate fitting balances model complexity and generalization, accurately representing the true relationship. Overfitting arises when the model is overly complex, fitting not only the true pattern but also noise in the training data, which harms predictive performance on unseen data.

Model training and validation typically involve three interconnected components:

- **Data splitting** to separate information for training, tuning, and final evaluation.

- **Learning objectives** that define what the model is optimizing for, such as minimizing forecasting error or maximizing classification accuracy.

- **Optimization techniques** that iteratively adjust model parameters to best satisfy the learning objectives.

- **Model validation** and evaluation to assess generalization performance, select hyperparameters, and ensure robustness before deployment.

These steps work together to maximize predictive accuracy while ensuring robustness against unseen conditions. The following subsections provide a structured overview of how models are trained, tuned, and validated in AD research.

4.3.1 Data Splitting

Before training, datasets must be divided into subsets for different purposes:
- Training set: Used to fit model parameters (e.g., weights in neural networks)
- Validation set: Used during training to tune hyperparameters and monitor overfitting
- Test set: Used only once, after training, to assess final model performance

Typical split ratios in AD applications are 60-70% of training set, 15-20% validationand 15-20% for test set. For time series data, chronological splitting is normally used to avoid information leakage which means training on future data.

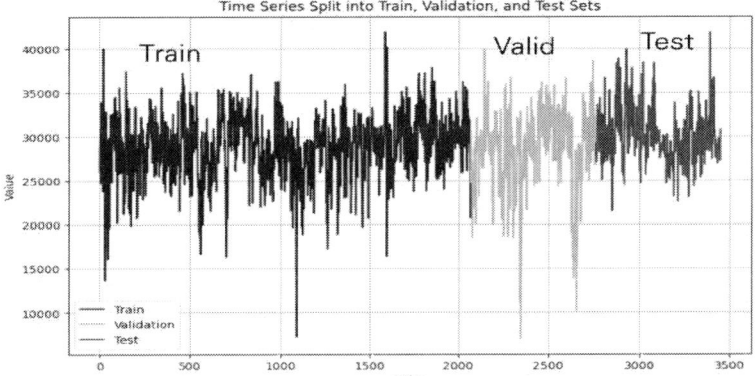

Fig. 4.19 Chronological splitting of time series data (Splitting ratio 6:2:2 for train, valid, test)

4.3.2 Learning Objectives

Learning objectives define what the model aims to minimize (loss function). The choice of loss function depends on the task:

- Regression tasks (e.g. predicting methane yield or biogas volume):

$$\text{MSE} = \frac{1}{N}\sum_{i=1}^{N}(y_i - \hat{y}_i)^2 \quad \ldots\ldots\ldots\ldots\ldots\ldots\ldots\ldots (4.40)$$

$$\text{MAE} = \frac{1}{N}\sum_{i=1}^{N}|y_i - \hat{y}_i| \quad \ldots\ldots\ldots\ldots\ldots\ldots\ldots\ldots (4.41)$$

Where y_i is the true value, $\hat{y}_i$ is the predicted value, and N is the number of samples.

② Classification tasks (e.g. stable vs unstable state classification):

$$\text{Cross-Entropy Loss} = -\sum_{i=1}^{N}\sum_{c=1}^{C} y_{i,c} \log(\hat{y}_{i,c}) \quad \cdots \cdots \cdots (4.42)$$

where C is the number of classes, $y_{i,c}$ is a binary indicator (1 if sample i belongs to class c, 0 otherwise), and $\hat{y}_{i,c}$ is the predicted probability for class c.

③ Anomaly Detection: Reconstruction Loss (e.g. autoencoder MSE)

$$\text{Reconstruction Loss} = \frac{1}{N}\sum_{i=1}^{N} \|x_i - \hat{x}_i\|^2 \quad \cdots \cdots \cdots (4.43)$$

where x_i is the original input and $\hat{x}_i$ is its reconstruction by the model. High reconstruction error often indicates anomalous conditions.

4.3.3 Model Optimization

Optimization techniques lie at the heart of model training, as they define how amodel adjusts its parameters to minimize the chosen loss function. Two key components govern this process: (1) the calculation of gradients, which measure how much each parameter contributes to the error, and (2) the update strategy, which determines how these gradients are used to refine parameters.

Backpropagation (Gradient Calculation)

Backpropagation is the standard algorithm used in training neural networks to compute gradients efficiently. The procedure involves two phases:

① **Forward pass**: Input data are propagated through the network to generate predictions.

① **Backward pass**: The loss function, which measures prediction error, is differentiated with respect to model parameters using the chain rule of calculus.

This "backpropagates" error signals layer bylayer, yielding gradients for each parameter.
Mathematically, for a parameter θ, the gradient of the loss L is expressed as:

$$\frac{\partial L}{\partial \theta} \quad \ldots\ldots\ldots\ldots\ldots\ldots\ldots\ldots\ldots\ldots\ldots (4.44)$$

These gradients provide the necessary information about how to adjust weights and biases to reduce the error. Without backpropagation, the training of deep networks with thousands or millions of parameters would be computationally infeasible.

Gradient Descent and Variants (Parameter Update Strategies)

Once gradients are computed, optimization algorithms use them to update parameters in a direction that reduces the loss. The most basic form is **gradient descent**, which updates parameters iteratively as:

$$\theta_{t+1} = \theta_t - \eta \nabla_\theta \quad \ldots\ldots\ldots\ldots\ldots\ldots\ldots\ldots (4.45)$$

where θ_t denotes the parameter at iteration t, η is the learning rate, and ∇_θ is the gradient of the loss. Several variants improve upon basic gradient descent, particularly for noisy, high-dimensional data commonly found in AD research:

- **Stochastic Gradient Descent (SGD):** Updates parameters using a single (or small batch of) data sample(s), enabling faster iterations and better scalability.

- **Mini-Batch Gradient Descent:** Strikes a balance by updating parameters using small batches of data (e.g., 32 or 64 samples). This reduces variance in gradient estimates compared to SGD, improves computational efficiency, and allows parallelization on GPUs. Mini-batching is the dominant approach in modern deep learning.

- **Momentum**: Adds a fraction of the previous update to the current step, helping the model escape shallow local minima.

- **RMSProp**: Adjusts the learning rate for each parameter adaptively based on the moving average of squared gradients.

- **Adam (Adaptive Moment Estimation)**: Combines the benefits of momentum and RMSProp, maintaining both running averages of gradients and squared gradients, making it one of the most widely used optimizers.

In practice, mini-batch training with adaptive optimizers such as Adam is thedefault choice for deep learning models, including those applied in AD research, as it balances stability, efficiency, and generalization performance.

4.3.4 Validation and Evaluation of Models

While training determines how well a model fits data, validation and evaluation determine how well it *generalizes* to new, unseen conditions. In AD research, this step is essential because models must perform reliably under variable feedstocks, operating regimes, and environmental conditions. A model that only performs well on training data but fails on unseen cases is not useful for process monitoring or decision support.

Cross-Validation Strategies

Validation is typically achieved by splitting data into training, validation, and test sets. However, when data are limited (as is often the case in AD research), **cross-validation** techniques are used to maximize data utilization:

- **k-Fold Cross-Validation:** Data are partitioned into k folds. The model is trained on $k-1$ folds and validated on the remaining fold, repeated k times.

- **Leave-One-Out Cross-Validation (LOOCV):** Each sample is used once as a validation point, useful for very small datasets.

- **Time Series Cross-Validation:** In AD applications, chronological order must be preserved to prevent information leakage. Here, training sets are progressively expanded, and validation is performed on subsequent time windows.

These strategies help detect overfitting, tune hyperparameters, and estimate generalization performance.

Evaluation Metrics

Evaluation metrics quantify model performance in alignment with the task type. Common metrics include:

- **Regression Tasks**

$$\text{RMSE} = \sqrt{\frac{1}{N}\sum_{i=1}^{N}(y_i - \hat{y}_i)^2} \quad \ldots\ldots\ldots\ldots\ldots\ldots (4.46)$$

$$R^2 = 1 - \frac{\sum_{i=1}^{N}(y_i - \hat{y}_i)^2}{\sum_{i=1}^{N}(y_i - \bar{y})^2} \quad \text{(4.47)}$$

Where RMSE measures average error magnitude, and R^2 indicates how much variance in the data is explained by the model.

Classification Tasks

$$\text{Accuracy} = \frac{TP + TN}{TP + TN + FP + FN} \quad \text{(4.48)}$$

$$\text{Precision} = \frac{TP}{TP + FP}, \quad \text{Recall} = \frac{TP}{TP + FN} \quad \text{(4.49)}$$

$$F1 = 2 \cdot \frac{\text{Precision} \cdot \text{Recall}}{\text{Precision} + \text{Recall}} \quad \text{(4.50)}$$

where TP, TN, FP, FN represent true positives, true negatives, false positives, and false negatives. F1 is particularly valuable in AD research where class imbalance (few instability cases vs. many stable cases) is common.

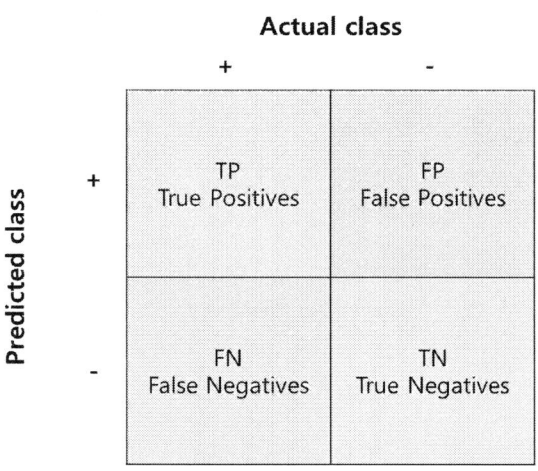

Fig. 4.20 Confusion matrix for binary classification

- **Anomaly Detection Tasks**

Models are evaluated using reconstruction error thresholds or statistical measures such as the Area Under the Receiver Operating Characteristic Curve (AUROC):

$$\mathrm{AUROC} = \int_0^1 \mathrm{TPR(FPR)}\, d(\mathrm{FPR}) \quad \ldots\ldots\ldots\ldots\ldots\ldots\ldots\ldots (4.51)$$

where TPR is the true positive rate and FPR is the false positive rate. High AUROC indicates good separation between normal and anomalous states.

Model validation and evaluation ensure that AI-driven insights in AD are trustworthy and applicable across different systems. By systematically checking performance with proper metrics and validation schemes, researchers avoid misleading conclusions and ensure that models can be safely deployed in process monitoring, forecasting, or control tasks.

Chapter 5. Application of AI in Anaerobic Digestion research

AD systems produce diverse datasets encompassing physicochemical sensor measurements, microbial community profiles, and imaging-derived observations. A clear understanding of these data types, their collection methods, and their interpretive value is essential for the effective application of AItechniques in monitoring, control, and optimization. The datasets described in this chapter have been, or can potentially be, applied in the development and training of AI models.

5.1 Physico-Chemical Data

Physico-chemical data forms the foundation of monitoring and analysis in anaerobic digestion (AD) systems. These data reflect the operational status, substrate characteristics, degradation performance, and gas production efficiency of the reactors. Collected at various points in the system — from input to effluent and gas phase —this data is indispensable for AI-based modeling, control, and forecasting applications.

5.1.1 Reactor Configuration

The performance of an AD system is largely influenced by its physical configuration and operational parameters. These include:

- **Influent Flow Rate (Q)**: Measured in L/day or m³/day, this indicates the volume of substrate entering the reactor, influencing hydraulic retention time and loading rates.
- **Hydraulic Retention Time (HRT):** Calculated as the ratio of reactor working volume to influent flow rate (typically in days). It determines how long the substrate remains in the digester, affecting degradation extent and gas yield.
- **Working Volume:** The active volume of the reactor (m³) where digestion occurs, excluding headspace.

- **Reactor Type:** Configuration may vary from continuous stirred-tank reactors (CSTRs) to plug-flow or upflow anaerobic sludge blanket (UASB) systems, each introducing distinct hydrodynamic and mixing characteristics.
- **Temperature Regime:** Mesophilic (30–40°C) or thermophilic (50–60°C) operation influences microbial community structure and degradation rates.

These variables often act as contextual or static inputs to AI models, helping interpret dynamic time-series data within an operational frame.

5.1.2 Substrate Characteristics

The composition of the feedstock is a critical determinant of digestion performance. Substrate data include:

- **Substrate Type:** Food waste, manure, sludge, energy crops, co-digested mixtures, or industrial byproducts. Substrate heterogeneity affects digestion stability and methane yield.

- **Substrate Volume:** The amount of substrate fed daily, typically in liters or m³, used to compute organic loading rate (OLR).
- **pH:** Indicates the acidity/alkalinity of the substrate; extreme values can inhibit microbial activity.
- **Chemical Oxygen Demand (COD):** Total oxidizable organic matter, often in mg/L or g/L. It provides a theoretical estimate of biogas production potential.
- **Solids:**
  - **Total Solids (TS):** Overall solid content (%, g/L).
  - **Volatile Solids (VS):** Organic fraction of the TS that can be degraded anaerobically.
- **Alkalinity:** Buffering capacity of the feedstock (usually in mg/L $CaCO_3$), crucial for pH stabilization during acid production.

AI models often use these values to predict downstream parameters like VFA accumulation or methane yield.

5.1.3 Effluent Characteristics

Effluent data reflect the outcome of anaerobic digestion and are vital for evaluating performance, stability, and treatment efficiency.

- **pH:** A critical stability indicator; extreme drops may signal process failure or overload.
- **COD (Total and Soluble):** Residual organic matter post-digestion. High effluent COD suggests poor substrate conversion.
- **Solids:**
  - **TS, VS:** Remaining total and volatile solids in the digestate.
  - **TSS/VSS:** Total and volatile suspended solids, often used in wastewater-based digesters.
- **Alkalinity:** Maintained by microbial activity; decreasing trends may warn of VFA accumulation.

- **Volatile Fatty Acids (VFA):** Short-chain fatty acids (e.g., acetate, propionate, butyrate), whose accumulation indicates process imbalance.
- **Ions:** Monitoring of NH_4^+, K^+, Mg^{2+}, and Na^+ is essential due to potential inhibitory effects at high concentrations.
- **Total Carbohydrates, Proteins, Lipids:** These reflect residual biodegradable fractions and help estimate degradation efficiency by macromolecule class.

In AI workflows, time-series of effluent parameters are used either as features or targets in prediction tasks or indicators in classification models (e.g., stable vs. unstable states).

5.1.4 Gas Phase Data

Gas production data directly reflects the functional output of the AD process and is key to evaluating reactor efficiency and energy recovery.

- **Gas Volume:** Measured daily (L/day or m³/day). Models often forecast this to monitor energy output or detect inhibition early.
- **Composition:**
  - **Methane (CH₄):** The primary energy product; typically, 50–70% of biogas volume.
  - **Carbon Dioxide (CO₂):** Secondary gas; high levels can indicate incomplete degradation or low methanogenic activity.
  - **Hydrogen Sulfide (H₂S):** A byproduct of sulfur-containing compounds; corrosive and toxic, even at low concentrations (ppm level).
  - **Ammonia (NH₃/NH₄⁺):** Produced during protein degradation; toxic to methanogens at high concentrations and often monitored both in liquid and gas phases.

Gas parameters serve both as **model targets** (for regression tasks) and **features** in multi-variate diagnostics. For instance, a drop in methane concentration coupled with a rise in H₂S may suggest substrate imbalance or microbial inhibition.

5.1.5 Application in AI Modeling

Numerical features in AD datasets generally require basic preprocessing before modeling. Physicochemical variables in AD come from heterogeneousinstruments (online sensors, lab assays) and arrive on different scales and units. Preprocessing make these data numerically comparable, statistically well-behaved, and safe to feed into learning algorithms.

1) Comparable scales for learning

Apply standardization (zero mean, unit variance) or min–max scaling to improve optimization and distance computations in SVMs, neural networks, and tree-adjacent methods that use distances in kernels or metrics. Keep the scaler fitted on the training split only to avoid leakage; transform validation/test with the same fitted scaler.

2) Variance-stabilizing transforms

Many concentration variables are skewed and heteroscedastic. Use power transforms such as Box-Cox (positive values only) or Yeo-Johnson (supports non-positive values) to stabilize variance and improve linear separability.

3) Robust scaling for outliers

When sensor spikes or lab anomalies are present, RobustScaler (median/IQR) is less sensitive than mean-std scaling. As with all preprocessing, fit on the training data only.

5.2 Qualitative and Quantitative Microbial Data

The performance of anaerobic digestion (AD) is intrinsically linked to the metabolic activity and ecological dynamics of its microbial communities. Bacteria and archaea drive the biochemical cascade—hydrolysis, acidogenesis, acetogenesis, and methanogenesis—that converts organic matter into methane-rich biogas. Understanding "who is there," "how much," and "how they change" provides early warning of instability, supports mechanism-aware control, and enables data-driven optimization. Microbial data in AD fall broadly into two complementary classes:

- Qualitative (compositional) data — community structure and diversity (typically from 16S rRNA gene amplicon sequencing or metagenomics);
- Quantitative data — absolute abundances of target groups, often via qPCR/dPCR of functional gene.

Integrating both views with process metadata (e.g., pH, alkalinity, VFAs) yields the richest insight for modeling and control. Large cross-plant surveys now map global AD microbiomes, providing "priors" on core taxa and ecological guilds that can guide feature engineering for AI models.

5.2.1 Qualitative Microbial Data

Qualitative profiles are derived chiefly from 16S rRNA gene sequencing (increasingly full-length reads), producing amplicon sequence variants (ASVs) or operational taxonomic units (OTUs) and corresponding taxonomic assignments. These data answer "who is present" and "in what proportion," not absolute counts. Full-scale surveys across hundreds of digesters report characteristic bacterial and archaeal consortia that recur across geographies and substrates.

5.2.1.1 Community Structure

- **Relative Abundance:** The proportion of each taxon (e.g., phylum, genus) within a sample, expressed as a percentage. For example, monitoring the proportion of methanogens like *Methanosaeta* or *Methanobacterium* via 16S rRNA sequencing is a routine approach to assess process stability and metabolic functioning (Dueholm et al., 2024)

- **Taxonomic hierarchy:** Reads are classified into Domain (Bacteria/Archaea), Phylum (e.g., Firmicutes, Bacteroidetes, Euryarchaeota), Class, Order, Family, Genus, and Species. Aggregation to higher ranks (e.g., family/genus) is common to stabilize features for learning while retaining ecological interpretability.
- Heat maps and bar plots are commonly used to visualize the distribution of taxa across samples or time. Interaction in microbial community dynamics are also represented through network analysis.

5.2.1.2 Diversity Indices

Microbial diversity reflects community stability, resilience, and functional redundancy.

- **Alpha Diversity (within-sample diversity):**

- Shannon Index: Accounts for both richness and evenness. Higher values suggest a well-balanced microbial ecosystem.
- Chao1: Estimates the total number of taxa, including those under-sampled.

- **Beta Diversity (between-sample diversity):**
  - Quantifies community dissimilarity using metrics like Bray-Curtis or UniFrac distances.
  - Often visualized via Principal Coordinates Analysis (PCoA) or Non-Metric Multidimensional Scaling (NMDS).
  - Phylogenetic Diversity: Incorporates evolutionary relationships between taxa, offering functional interpretability.

5.2.1.3 Applications in AI Modeling

Amplicon-based relative abundances are compositional which means values carry only relative information, constrained to a constant sum. Standard statistical/ML methods can mislead unless we respect this property. Treating microbiome data as compositions is now widely recommend to avoid spurious correlations and improve inference.

Qualitative microbial profiles carry invaluable clues about reactor stability, yet they pose a classic "large-p, small-n"problem for AI: each sample may contain on the order of ~200 genera (features) while only a few dozen time points are affordable to sequence. This high-dimensional, low-sample setting greatly amplifies the risk of overfitting—i.e., a model that captures noise specific to the training data instead of patterns that generalize to new reactors. To curb this, there are several complementary tactics:

- Dimensionality-reduction (e.g., principal component analysis, non-negative matrix factorization, variational auto-encoders) to project the community matrix onto a compact latent space.
- Feature-selection or sparsity-inducing regularization (LASSO, elastic-net, group LASSO) to keep only taxa that add predictive value.

- Hierarchicalpooling or taxonomic aggregation to roll rare species into higher-level clades, reducing noise while preserving ecological meaning.
- Data-augmentation and transfer-learning approaches—such as synthetic minority over-sampling (SMOTE) or pre-training on external microbiome datasets—to effectively enlarge the sample size and encode prior biological knowledge.

Used together with rigorous cross-validation, these strategies enable robust AI models that translate complex community-structure signals into actionable predictions despite limited sequencing budgets.

5.2.2 Quantitative Microbial Data

While qualitative microbial data describe community composition, quantitative microbial data provide measurements of the absolute abundance of specific microbial groups, commonly obtained using techniques such as quantitative PCR (qPCR).

5.2.2.1 Microbial Quantification using a qPCR

- **Target-specific quantification:** qPCR assays are designed to quantify functional microbial groups or key metabolic players, such as:
  - Hydrolytic bacteria (e.g., *Clostridium spp.*)
  - Acidogens (e.g., *Syntrophomonas*)
  - Acetogens (e.g., *Syntrophobacter*)
  - Methanogens (e.g. *Methanosaeta* (acetoclastic), *Methanobacterium* (hydrogenotrophic))
- **Units:** Gene copies per mL or gram (e.g., 1.2×10^7 gene copies/mL).
- **Time-resolved quantification:** Time series of qPCR data can reveal population dynamics during load shock, inhibition, or recovery phases.

5.2.2.2 Applications in AI Modeling

Although quantitative microbial datasets have been less frequently incorporated as features in AI models compared to qualitative data, they hold significant potential for enhancing model performance by offering direct, functionally relevant indicators of microbial activity. When combined with reactor performance data, these counts enable causal inference, helping to link microbial shifts to process outcomes. Since these datasets are typically lower in dimensionality and more interpretable, they are well suited for traditional machine learning models (e.g., decision trees, SVM).

5.2.3 Integration of Qualitative and Quantitative Data

An integrated approach—leveraging both 16S sequencing and qPCR—provides a comprehensive understanding of AD microbial ecology:
- Sequencing data reveal community composition and shifts.
- qPCR data confirm functional relevance by quantifying target populations.

For AI applications, combining these datasets enhances **feature richness** and improves **model**

interpretability, especially in hybrid models where biological knowledge supports data-driven predictions. These are some of the examples where microbial data can be used in AI modeling.

- **Biogas Production Prediction**

  Supervised models (e.g., tree ensembles, kernels, and deep regressors) trained on microbial profiles often fused with physicochemical covariates, have reported high predictive performance for biogas yield and methane content, with several case studies noting R^2 values > 0.8 (Dang et al., 2025; Li et al., 2022; Long et al., 2021; Yu et al., 2024; Zhang et al., 2023).

- **Process Optimization**

  ML has been used to identify operating windows and microbiome configurations associated with stable, high-yield regimes. For example, studies have highlighted beneficial ranges in the relative abundance of key methanogens—such as Methanoculleus ($\approx$15–37%) and Methanomicrobiaceae ($\approx$3–15%)—as stability markers to target during optimization (Yu et al., 2024).

- **Failure Prediction and System Monitoring**

    Early-warning classifiers that track coordinated shifts in microbial composition alongside physicochemical trendscan flag impending instability (e.g., VFA accumulation, pathway shifts) ahead of performance collapse, enabling proactive control (Vanwonterghem et al., 2014).

- **Microbial Biomarker Identification**

    Feature-attribution analyses frequently elevate taxa with known functional role (e.g., Methanosarcina, Methanoculleus, and syntrophic partner) as biomarkers of performance, strengthening mechanistic interpretability and guiding targeted monitoring (Liu et al., 2022; Piercy et al., 2025)

5.3 Image Data

Image-based data represent an emerging and powerful resource in AD research, providing a non-invasive, high-resolution view of both substrate characteristics and microbial dynamics. Unlike traditional numerical datasets, images capture spatial patterns, morphological structures, and spectral signatures that are often difficult to quantify through conventional measurements. A notable advantage is their capacity to generate large volumes of information far beyond what can be processed manually, thereby overcoming the labor-intensive nature of traditional analyses. With advancements in AI—particularly computer vision—image data are increasingly leveraged for automated monitoring, quality assessment, and microbialcharacterization in AD systems.

Image data in AD can be broadly categorized into two major types:

- **Spectroscopy image:** Derived from the analysis of reflected or transmitted light through substrates, often used to predict chemical composition.
- **Microscopy image:** High-magnification visuals used to observe microbial communities and biofilm formation.

5.3.1 Spectroscopy Image

Spectroscopy-based imaging techniques, such as **hyperspectral imaging (HSI)** or **near-infrared (NIR) imaging**, capture data across a wide range of wavelengths. These images are often used to analyze the chemical composition of influents and effluents without the need for wet-lab analysis.

5.3.1.1 Application in Influent Analysis

Spectroscopic imaging of the incoming feedstock can be used to:
- Estimate total solids, COD, protein, or lipid content based on spectral reflectance.
- Differentiate between feedstock types, such as food waste vs. manure, based on spectral fingerprints.
- Detect contamination or variation in substrate composition in real time.

5.3.1.2 Application in Effluent Analysis

For the effluent, spectroscopy images help:

- Predict residual COD or TS/VS levels, supporting evaluation of treatment efficiency.
- Identify undigested organic materials or inhibitory compounds.
- Assess color and turbidity, which correlate with effluent quality and stability.

5.3.1.3 Applications in AI Modeling
- AI models (e.g., CNNs or spectral regression networks) can be trained to predict physicochemical properties directly from pixel-level spectral information.
- Dimensionality reduction techniques such as PCA or t-SNE are commonly applied to handle the high-dimensional nature of hyperspectral images.

Several peer-reviewed studies have demonstrated the integration of spectral imaging with machine learning for AD monitoring:
- Peng et al., (2024, Environmental Science & Technology) developed a near-infrared HSI

workflow to predict digestate parameters (total nitrogen (TN), total organic carbon (TOC), total ammonia nitrogen (TAN), and COD) from spectral data, and to detect floating granules in lipid-rich reactors via target-detection algorithms. Using 732 digestate samples, dried-digestate prediction models achieved R²_test = 0.86 (RMSE = 690 mg L⁻¹) for TN and R²_test = 0.82 (RMSE = 1090 mg L⁻¹) for TOC.

- Serranti et al. (2012, Optical Engineering) applied NIR-HSI (1000–1700 nm) to monitor wine-waste anaerobic digestion. Spectral data were modeled with partial least squares (PLS) regression, achieving calibration R^2 up to 0.90 for COD (RMSECV ≈ 5495 mg O_2 L⁻¹) and R^2 = 0.68 for nitrate (RMSECV ≈ 12.8 mg L⁻¹).
- Peng et al. (2022, Renewable & Sustainable Energy Reviews) reviewed NIR spectroscopy and hyperspectral imaging applications in AD, highlighting their coupling with chemometric and

machine-learning methods for feedstock characterization, digestate quality monitoring, and early detection of process upsets.
- Neurauter et al. (2024, FEMS Microbes) demonstrated the use of NIR spectroscopy and HSI for nucleic-acid-independent characterization of anaerobic gut fungi strains relevant to anaerobic systems, achieving successful strain discrimination through discriminant analysis models.

5.3.2 Microscopy Image

Microscopic imaging provides a close-up view of the microbial and structural elements within the digester, particularly in the effluent. This includes observations of:
- Microbial cells and aggregates
- Biofilm development
- Granular sludge morphology

- Presence of filamentous bacteria (often linked to foaming or bulking)

5.3.2.1 Biofilm and Granule Analysis

- In systems like up flow anaerobic sludge blanket (UASB) reactors, granular sludge morphology is critical for process performance.
- Microscopy allows researchers to assess granule size, shape, and integrity, which can correlate with stability and methanogenic capacity.
- Confocal laser scanning microscopy (CLSM) and scanning electron microscopy (SEM) can provide structural details of biofilms and microbial attachment.

5.3.2.2 Microbial Community Imaging

- Staining techniques (e.g., DAPI, FISH) are used to label specific microbial groups, allowing visualization of methanogens, acidogens, or syntrophic associations.

- Image analysis can quantify:
    - Cell density and distribution
    - Community clustering and spatial heterogeneity
    - Microbial morphological traits (e.g., cocci, rods, filaments)

5.3.2.3 Applications in AI Modeling

AI-based image processing techniques—especially convolutional neural networks (CNNs)—can automate the analysis of microscopy images:
- Cell segmentation and counting
- Classification of microbial morphotypes
- Detection of abnormal structures (e.g., biofilm detachment or damaged granules)

These models help reduce manual labor, improve consistency, and enable real-time diagnostic capabilities for monitoring digester health. While the application of microscopy image data for AI model development in anaerobic digestion has not yet been reported, analogous work has been conducted in wastewater treatment. Borzooei et al. (2024, Journal of Water Process Engineering) compiled a two-year dataset of weekly microscopic images and employed deep convolutional neural networks (Inception V3, ResNet18, ResNet152, ConvNeXt-nano, and ConvNeXt-S) with transfer learning to predict sludge volume index, a key settling performance metric. The models provided objective, consistent, and less labor-intensive evaluations than manual analysis, enabling real-time applicability. These findings suggest that microscopy-based AI approaches could be adapted to predict operational states in anaerobic digesters.

5.4 Time-series Data

Time-series datasets are a central data type in AD research, representing sequential measurements collected over time. They preserve the temporal order of observations, allowing for the analysis of dynamic patterns such as trends, seasonal variations, and short-term disturbances. In AD operations, most monitoring parameters whether from continuous sensors, periodic laboratory analyses, or automated control systems, are recorded as time-series, making them indispensable for process control and AI-based forecasting.

In addition to standard operational and quality measurements, time-series data in AD often include event or incident logs. These logs record discrete occurrences that may influence process behavior but are not continuous measurements, such as:
- Feedstock change events (e.g., switching from food waste to manure)
- Maintenance activities or cleaning operations

- Unplanned shutdowns or start-ups
- Temperature shifts outside the normal range
- Episodes of foaming, scum accumulation, or inhibition events

These event annotations provide valuable context for interpreting anomalies or shifts in time-series patterns, particularly in root-cause analysis or supervised learning where event labels can serve as classification targets.

5.4.1 Characteristics of Time-series Data

- Serial correlation: Measurements are temporally dependent, meaning current values are influenced by previous states.
- Multivariate structure: Dozens of interrelated variables are recorded simultaneously, requiring joint analysis.
- Non-stationarity: Statistical properties such as mean and variance may change over time due to

seasonal effects, operational adjustments, or process degradation.
- Multiple time scales: Variables may vary on hourly, daily, or seasonal scales, introducing complexity in model design.
- Missing or irregular sampling: Sensor malfunctions, maintenance downtime, or manual sampling schedules can introduce gaps or uneven intervals.

5.4.2 Preprocessing Requirements for AI Applications

To prepare AD time-series datasets for statistical or machine learning models, the following preprocessing steps are commonly applied:
- Resampling to uniform intervals to ensure consistent time steps across variables.
- Interpolation or imputation of missing values using methods such as linear interpolation, spline fitting, or forward/backward filling.
- Normalization or standardization to remove scale differences between variables.

- Detrending or differencing for stationary model requirements, especially in forecasting tasks.
- Outlier detection and correction to mitigate the impact of sensor faults or recording errors.
- Encoding event logs into binary or categorical features for integration into predictive models.

Proper preprocessing not only improves model performance but also ensures that learned patterns reflect true process dynamics rather than artifacts of data collection or irregular sampling.

5.4.3 Applications in AI Modeling

Below are several notable, peer-reviewed journal studies that applied time-series datasets in machine learning or AI models for AD:

- Sappl et al. (2023, Science of total environment) applied Temporal Fusion Transformer model using six years of full-scale operational time-series data, enriched with categorical features (e.g., public holidays), to forecast biogas production. The model outputs include quantiles (not just median forecasts), improving robustness to fluctuations. Forecasting accuracy (Mean Absolute Percentage Error) was below 8%.

- Schroer & Just (2024, ES&T engineering) used minute-scale SCADA data from a municipal co-digestion facility to forecast biogas production with an MLP neural network. The model achieved an R^2 of 0.78 and a Mean Absolute Percentage Error of 13.4 % on hold-out data, despite the addition of daily lab data offering minimal performance gains.
- (Han et al., 2025) implemented an iTransformer-based time-series model to forecast biogas production from food waste AD, achieving high short-term predictive accuracy.

References

Adekunle, K.F., Okolie, J.A.J.A.i.B., Biotechnology. 2015. A review of biochemical process of anaerobic digestion. 6(03), 205.

Agyeman, F.O., Han, Y., Tao, W.J.B.t. 2021. Elucidating the kinetics of ammonia inhibition to anaerobic digestion through extended batch experiments and stimulation-inhibition modeling. 340, 125744.

Alzate Ibañz, A.M., Ocampo-Martinez, C., Cardona Alzate, C.A., Trejos M, V.M.J.a.e.-p. 2017. Monitoring management criterion of an anaerobic digester using an explicit model based on temperature and pH. arXiv: 1706.02187.

Anderson, M.J.J.A.e. 2001. A new method for non-parametric multivariate analysis of variance. 26(1), 32-46.

Appels, L., Baeyens, J., Degrèe, J., Dewil, R.J.P.i.e., science, c. 2008. Principles and potential of the anaerobic digestion of waste-activated sludge. 34(6), 755-781.

Bakker, J.D.J.A.M.S.i.R. 2024. Nmds.

Batstone, D.J., Keller, J., Angelidaki, I., Kalyuzhnyi, S.V., Pavlostathis, S.G., Rozzi, A., Sanders, W., Siegrist, H., Vavilin, V.A.J.W.S., technology. 2002. The IWA anaerobic digestion model no 1 (ADM1). 45(10), 65-73.

Borcard, D., Gillet, F., Legendre, P. 2011. *Numerical ecology with R*. Springer.

Calusinska, M., Goux, X., Fosséré M., Muller, E.E., Wilmes, P., Delfosse, P.J.B.f.b. 2018. A year of monitoring 20 mesophilic full-scale bioreactors reveals the existence of stable but different core microbiomes in bio-waste and wastewater anaerobic digestion systems. 11(1), 196.

Caporaso, J.G., Kuczynski, J., Stombaugh, J., Bittinger, K., Bushman, F.D., Costello, E.K., Fierer, N., Peñ, A.G., Goodrich, J.K., Gordon, J.I.J.N.m. 2010. QIIME allows analysis of high-throughput community sequencing data. 7(5), 335-336.

Cho, K., Lee, J., Kim, W., Hwang, S.J.P.B. 2013. Behavior of methanogens during start-up of farm-scale anaerobic digester treating swine wastewater. 48(9), 1441-1445.

Crowley, P.H., Straley, S.C., Craig, R.J., Culin, J.D., Fu, Y.T., Hayden, T.L., Robinson, T.A., Straley, J.P.J.J.o.T.B. 1980. A model of prey bacteria, predator bacteria, and bacteriophage in continuous culture. 86(2), 377-400.

Dai, X., Yan, H., Li, N., He, J., Ding, Y., Dai, L., Dong, B.J.S.r. 2016. Metabolic adaptation of microbial communities to ammonium stress in a high solid anaerobic digester with dewatered sludge. 6(1), 28193.

Demirel, B., Scherer, P.J.R.i.E.S., Bio/Technology. 2008. The roles of acetotrophic and hydrogenotrophic methanogens during anaerobic conversion of biomass to methane: a review. 7(2), 173-190.

Draper, N. 1998. *Applied regression analysis*. McGraw-Hill. Inc.

El-Mashad, H.M., Zhang, R.J.B.t. 2010. Biogas production from co-digestion of dairy manure and food waste. 101(11), 4021-4028.

Elpianora, E., Berou, M., Kong, X., Hun, K., Azadegan, E. 2024. Fourth order runge-kutta and gill methods in numerical analysis of predator-prey models. Interval: Indonesian Journal of Mathematical Education, 2 (2), 164-177.

Faith, D.P., Minchin, P.R., Belbin, L.J.V. 1987. Compositional dissimilarity as a robust measure of ecological distance. 69(1), 57-68.

Ferreira, S.C., Bruns, R., Ferreira, H.S., Matos, G.D., David, J., Brandã, G., da Silva, E.P., Portugal, L., Dos Reis, P., Souza, A.J.A.c.a. 2007. Box-Behnken design: An alternative for the optimization of analytical methods. 597(2), 179-186.

Gaby, J.C., Zamanzadeh, M., Horn, S.J.J.B.f.b. 2017. The effect of temperature and retention time on methane production and microbial community composition in staged anaerobic digesters fed with food waste. 10(1), 302.

Gerber, M., Span, R.J.P.I., Paris. 2008. An analysis of available mathematical models for anaerobic digestion of organic substances for production of biogas.

Ghofrani-Isfahani, P., Valverde-Péez, B., Alvarado-Morales, M., Shahrokhi, M., Vossoughi, M., Angelidaki, I.J.C.E.J. 2020. Supervisory control of an anaerobic digester subject to drastic substrate changes. 391, 123502.

Goux, X., Calusinska, M., Lemaigre, S., Marynowska, M., Klocke, M., Udelhoven, T., Benizri, E., Delfosse, P.J.B.f.b. 2015. Microbial community dynamics in replicate anaerobic digesters exposed sequentially to increasing organic loading rate, acidosis, and process recovery. 8(1), 122.

Gundersen, M.S., Morelan, I.A., Andersen, T., Bakke, I., Vadstein, O.J.I.c. 2021. The effect of periodic disturbances and carrying capacity on the significance of selection and drift in complex bacterial communities. 1(1), 53.

Guseva, K., Darcy, S., Simon, E., Alteio, L.V., Montesinos-Navarro, A., Kaiser, C.J.S.B., Biochemistry. 2022. From diversity to complexity: Microbial networks in soils. 169, 108604.

Han, Y., Zeng, C., Ni, Q., Wang, J., Chu, Z., Zhang, X., Geng, Z., Tan, L., Liu, Y.J.C.E.J. 2025. Time series prediction of anaerobic digestion yield and carbon emissions from food waste based on iTransformer model. 513, 163064.

Jannasch, H.W., Egli, T.J.A.v.L. 1993. Microbial growth kinetics: a historical perspective. 63(3), 213-224.

Jannat, M.A.H., Lee, J., Shin, S.G., Hwang, S.J.J.o.H.M. 2021. Long-term enrichment of anaerobic propionate-oxidizing consortia: Syntrophic culture development and growth optimization. 401, 123230.

Jannat, M.A.H., Park, S.H., Chairattanawat, C., Yulisa, A., Hwang, S.J.B.T. 2022. Effect of different microbial seeds on batch anaerobic digestion of fish waste. 349, 126834.

Jannat, M.A.H., Park, S.H., Hwang, S.J.B.T. 2024. Modeling interactions of Clostridium cadaveris and Clostridium sporogenes in anaerobic acidogenesis of glucose and peptone. 393, 130099.

Jia, R., Song, Y.-C., An, Z., Kim, K., Lee, C.-Y., Bae, B.-U.J.P. 2023. A New Comprehensive Indicator for Monitoring Anaerobic Digestion: A Principal Component Analysis Approach. 12(1), 59.

Jolliffe, I.T., Cadima, J.J.P.t.o.t.r.s.A.M., Physical, Sciences, E. 2016. Principal component analysis: a review and recent developments. 374(2065), 20150202.

Kim, E., Lee, J., Han, G., Hwang, S.J.B.t. 2018. Comprehensive analysis of microbial communities in full-scale mesophilic and thermophilic anaerobic digesters treating food waste-recycling wastewater. 259, 442-450.

Kong, D., Zhang, K., Liang, J., Gao, W., Du, L.J.M. 2019. Methanogenic community during the anaerobic digestion of different substrates and organic loading rates. 8(5), e00709.

Koo, T., Jannat, M.A.H., Hwang, S.J.J.o.M., Biotechnology. 2020. Biokinetics of protein degrading Clostridium cadaveris and Clostridium sporogenes in batch and continuous mode of operations. 30(4), 533.

Koo, T., Lee, J., Hwang, S.J.J.o.E.M. 2019. Development of an interspecies interaction model: an experiment on Clostridium cadaveris and Clostridium sporogenes under anaerobic condition. 237, 247-254.

KováováKovar, K., Egli, T.J.M., reviews, m.b. 1998. Growth kinetics of suspended microbial cells: from single-substrate-controlled growth to mixed-substrate kinetics. 62(3), 646-666.

Lee, C., Kim, J., Shin, S.G., Hwang, S.J.F.m.e. 2008. Monitoring bacterial and archaeal community shifts in a mesophilic anaerobic batch reactor treating a high-strength organic wastewater. 65(3), 544-554.

Lee, E., Cumberbatch, J., Wang, M., Zhang, Q.J.B.t. 2017a. Kinetic parameter estimation model for anaerobic co-digestion of waste activated sludge and microalgae. 228, 9-17.

Lee, J., Shin, S.G., Han, G., Koo, T., Hwang, S.J.B.t. 2017b. Bacteria and archaea communities in full-scale thermophilic and mesophilic anaerobic digesters treating food wastewater: Key process parameters and microbial indicators of process instability. 245, 689-697.

Legendre, P., Legendre, L. 2012. *Numerical ecology*. Elsevier.

Leng, L., Yang, P., Singh, S., Zhuang, H., Xu, L., Chen, W.-H., Dolfing, J., Li, D., Zhang, Y., Zeng, H.J.B.t. 2018. A review on the bioenergetics of anaerobic microbial metabolism close to the thermodynamic limits and its implications for digestion applications. 247, 1095-1106.

Lever, J., Krzywinski, M., Altman, N.J.N.m. 2017. Points of significance: Principal component analysis. 14(7), 641-643.

Li, J., Rui, J., Yao, M., Zhang, S., Yan, X., Wang, Y., Yan, Z., Li, X.J.F.i.M. 2015. Substrate type and free ammonia determine bacterial community structure in full-scale mesophilic anaerobic digesters treating cattle or swine manure. 6, 1337.

Li, Y., Park, S.Y., Zhu, J.J.R., reviews, s.e. 2011. Solid-state anaerobic digestion for methane production from organic waste. 15(1), 821-826.

Liu, C., Wang, W., Anwar, N., Ma, Z., Liu, G., Zhang, R.J.E., fuels. 2017. Effect of organic loading rate on anaerobic digestion of food waste under mesophilic and thermophilic conditions. 31(3), 2976-2984.

Lovato, G., Kovalovszki, A., Alvarado-Morales, M., Jélot, A.T.A., Rodrigues, J.A.D., Angelidaki, I.J.R.E.

2021. Modelling bioaugmentation: Engineering intervention in anaerobic digestion. 175, 1080-1087.

Lozupone, C., Knight, R.J.A., microbiology, e. 2005. UniFrac: a new phylogenetic method for comparing microbial communities. 71(12), 8228-8235.

Maleki, E., Bokhary, A., Liao, B.J.R.i.E.S., Bio/Technology. 2018. A review of anaerobic digestion bio-kinetics. 17(4), 691-705.

Mata-Alvarez, J., Dosta, J., Macé S., Astals, S.J.C.r.i.b. 2011. Codigestion of solid wastes: a review of its uses and perspectives including modeling. 31(2), 99-111.

McMurdie, P.J., Holmes, S.J.P.o. 2013. phyloseq: an R package for reproducible interactive analysis and graphics of microbiome census data. 8(4), e61217.

Montgomery, D.C., Peck, E.A., Vining, G.G. 2021. *Introduction to linear regression analysis*. John Wiley & Sons.

Musa, H., Saidu, I., Waziri, M.J.I.J.o.C.A. 2010. A simplified derivation and analysis of fourth order Runge Kutta method. 9(8), 51-55.

Myers, R.H., Montgomery, D.C., Anderson-Cook, C.M. 2016. *Response surface methodology: process and product optimization using designed experiments*. John Wiley & Sons.

Neter, J., Kutner, M.H., Nachtsheim, C.J., Wasserman, W. 1996. Applied linear statistical models.

Oksanen, J.J. 2022. _vegan: Community Ecology Package_. R package version 2.6–4.

Pramanik, S.K., Suja, F.B., Porhemmat, M., Pramanik, B.K.J.P. 2019. Performance and kinetic model of a single-stage anaerobic digestion system operated at different successive operating stages for the treatment of food waste. 7(9), 600.

Rajagopal, R., Massé D.I., Singh, G.J.B.t. 2013. A critical review on inhibition of anaerobic digestion process by excess ammonia. 143, 632-641.

Ramette, A.J.F.m.e. 2007. Multivariate analyses in microbial ecology. 62(2), 142-160.

Rincó, B., Borja, R., Gonzáez, J., Portillo, M., Sáz-Jiméez, C.J.B.e.j. 2008. Influence of organic loading rate and hydraulic retention time on the performance, stability and microbial communities of one-stage anaerobic digestion of two-phase olive mill solid residue. 40(2), 253-261.

Schloss, P.D., Westcott, S.L., Ryabin, T., Hall, J.R., Hartmann, M., Hollister, E.B., Lesniewski, R.A., Oakley, B.B., Parks, D.H., Robinson, C.J.J.A., microbiology, e. 2009. Introducing mothur: open-source, platform-independent, community-supported software for describing and comparing microbial communities. 75(23), 7537-7541.

Schroer, H.W., Just, C.L.J.A.E., Engineering, t. 2023. Feature engineering and supervised machine learning to forecast biogas production during municipal anaerobic co-digestion. 4(3), 660-672.

Sundberg, C., Al-Soud, W.A., Larsson, M., Alm, E., Yekta, S.S., Svensson, B.H., Sørensen, S.J., Karlsson, A.J.F.m.e. 2013. 454 pyrosequencing analyses of bacterial and archaeal richness in 21 full-scale biogas digesters. 85(3), 612-626.

Ter Braak, C.J., Verdonschot, P.F.J.A.s. 1995. Canonical correspondence analysis and related multivariate methods in aquatic ecology. 57(3), 255-289.

Vanwonterghem, I., Jensen, P.D., Ho, D.P., Batstone, D.J., Tyson, G.W.J.C.o.i.b. 2014. Linking microbial community structure, interactions and function in anaerobic digesters using new molecular techniques. 27, 55-64.

Venkiteshwaran, K., Bocher, B., Maki, J., Zitomer, D.J.M.i. 2015. Relating anaerobic digestion microbial community and process function: supplementary issue: water microbiology. 8, MBI. S33593.

Wainaina, S., Lukitawesa, Kumar Awasthi, M., Taherzadeh, M.J.J.B. 2019. Bioengineering of anaerobic digestion for volatile fatty acids, hydrogen or methane production: a critical review. 10(1), 437-458.

Ward, A.J., Hobbs, P.J., Holliman, P.J., Jones, D.L.J.B.t. 2008. Optimisation of the anaerobic digestion of agricultural resources. 99(17), 7928-7940.

Werner, J.J., Garcia, M.L., Perkins, S.D., Yarasheski, K.E., Smith, S.R., Muegge, B.D., Stadermann, F.J.,DeRito, C.M., Floss, C., Madsen, E.L.J.A., microbiology, e. 2014. Microbial community dynamics and stability during an ammonia-induced shift to syntrophic acetate oxidation. 80(11), 3375-3383.

Westerholm, M. 2012. *Biogas production through the syntrophic acetate-oxidising pathway.*

Westerholm, M., Isaksson, S., Lindsjö O.K., Schnüer, A.J.A.E. 2018. Microbial community adaptability to altered temperature conditions determines the potential for process optimisation in biogas production. 226, 838-848.

Wu, Y., Kovalovszki, A., Pan, J., Lin, C., Liu, H., Duan, N., Angelidaki, I.J.B.f.b. 2019. Early warning indicators for mesophilic anaerobic digestion of corn stalk: a combined experimental and simulation approach. 12(1), 106.

Xu, R., Yang, Z.-H., Zheng, Y., Liu, J.-B., Xiong, W.-P., Zhang, Y.-R., Lu, Y., Xue, W.-J., Fan, C.-Z.J.B.T. 2018. Organic loading rate and hydraulic retention time shape distinct ecological networks of anaerobic digestion related microbiome. 262, 184-193.

Yasin, N.H.M., Mumtaz, T., Hassan, M.A., Abd Rahman, N.A.J.J.o.e.m. 2013. Food waste and food processing waste for biohydrogen production: a review. 130, 375-385.

Zamanzadeh, M., Hagen, L.H., Svensson, K., Linjordet, R., Horn, S.J.J.S.r. 2017. Biogas production from food waste via co-digestion and digestion-effects on performance and microbial ecology. 7(1), 17664.

Zamanzadeh, M., Hagen, L.H., Svensson, K., Linjordet, R., Horn, S.J.J.W.R. 2016. Anaerobic digestion of food waste–effect of recirculation and temperature on performance and microbiology. 96, 246-254.

Ziels, R.M., Svensson, B.H., Sundberg, C., Larsson, M., Karlsson, A., Yekta, S.S.J.M.b. 2018. Microbial rRNA gene expression and co-occurrence profiles associate with biokinetics and elemental composition in full-scale anaerobic digesters. 11(4), 694-709.

Acknowledgement

This work was supported by the Korea Ministry of Environment as Waste to Energy-Recycling Human Resource Development Project [No. YL-WE-21-002].